Constraint Handling in Cohort Intelligence Algorithm

Advances in Metaheuristics

Series Editors:
Anand J. Kulkarni, *Institute of Artificial Intelligence, MIT World Peace University, Pune, India*
Patrick Siarry, *Universite Paris-Est Creteil, France*

Handbook of AI-based Metaheuristics
Edited by Anand J. Kulkarni and Patrick Siarry

Metaheuristic Algorithms in Industry 4.0
Edited by Pritesh Shah, Ravi Sekhar, Anand J. Kulkarni and Patrick Siarry

Constraint Handling in Cohort Intelligence Algorithm
Ishaan R. Kale, Anand J. Kulkarni

For more information about this series please visit: www.routledge.com/ Advances-in-Metaheuristics/book-series/AIM

Constraint Handling in Cohort Intelligence Algorithm

Ishaan R. Kale and Anand J. Kulkarni

CRC Press
Taylor & Francis Group
Boca Raton London New York

CRC Press is an imprint of the
Taylor & Francis Group, an **informa** business

First edition published 2022
by CRC Press
6000 Broken Sound Parkway NW, Suite 300, Boca Raton, FL 33487-2742

and by CRC Press
2 Park Square, Milton Park, Abingdon, Oxon OX14 4RN

CRC Press is an imprint of Taylor & Francis Group, LLC

© 2022 Ishaan R. Kale, Anand J. Kulkarni

Library of Congress Cataloging-in-Publication Data
Names: Kale, Ishaan R., author. | Kulkarni, Anand Jayant, author.
Title: Constraint handling in cohort intelligence algorithm / Ishaan R. Kale, Anand J. Kulkarni.
Description: Boca Raton: CRC Press, 2022. |
Series: Advances in metaheuristics | Includes bibliographical references and index.
Identifiers: LCCN 2021036526 | ISBN 9781032150758 (hardback) |
ISBN 9781032156576 (paperback) | ISBN 9781003245193 (ebook)
Subjects: LCSH: Artificial intelligence. | Computational intelligence. | Algorithms.
Classification: LCC Q342 .K35 2022 | DDC 006.3–dc23
LC record available at https://lccn.loc.gov/2021036526

ISBN: 978-1-032-15075-8 (hbk)
ISBN: 978-1-032-15657-6 (pbk)
ISBN: 978-1-003-24519-3 (ebk)

DOI: 10.1201/9781003245193

Typeset in Minion
by Newgen Publishing UK

Access the Support Material: http://www.routledge.com/9781032150758

Contents

CHAPTER 1

Introduction to Metaheuristic Algorithms

1.1 WHAT IS A METAHEURISTIC ALGORITHM?

Engineering problems from different domains such as mechanical design engineering, truss structural, manufacturing and combinatorial require various analytical and experimental processes in order to build a mathematical model. Such models include several parameters and variables that are interdependent, as well as linear and nonlinear constraints. Most of these problems have both discrete and mixed design variables. These variables have a limited search space, due to which the solution may get stuck in local minima. Such problems are difficult/cumbersome to solve using traditional gradient-based optimisation methods. Hence, the researchers are motivated to apply Artificial Intelligence (AI)-based stochastic optimisation techniques. However, these stochastic optimisation techniques can not necessarily solve all types of problems. Problem-specific algorithms need to be developed; these are referred to as *heuristic algorithms*. Heuristic algorithms require several modifications in order to make them appropriate for solving problems from different domain; these modified algorithms are referred to as *metaheuristic algorithms*. Metaheuristic algorithms such as the Genetic Algorithm (GA) (Goldberg, 1989), Particle Swarm Optimisation (PSO) algorithm (Eberhart and Kennedy, 1995) and

DOI: 10.1201/9781003245193-1

the Ant Colony Algorithm (ACO) (Colorni et al., 1991) have shown their applicability in solving problems from different areas. AI-based stochastic optimisation techniques are classified as follows: a) bio-inspired, b) swarm intelligence and c) physics and chemical system-inspired. These techniques are also referred to as nature-inspired techniques. The detailed classification of these techniques is illustrated in Figure 2.1 (taken from Kumar et al., 2018).

1.2 DESIGN VARIABLES

Design variables play an important role in the optimisation process. They are classified as follows: a) continuous variables, b) discrete variables and c) integer variables. Let x_i be any design variable to be selected from defined range bounds (where L_b is the lower bound and U_b is the upper bound). The design variable is referred to as continuous if any finite value must be selected from within the range bounds $L_b \leq x_i \leq U_b$. Discrete variables are selected from a finite set of values as follows: Let set A consists of finite/discrete values, e.g., set $A = \{0.1, 0.3, 0.5, 0.7, 0.9\}$, within which the feasible design variables are selected to minimise/maximise the cost function. Such variables are referred to as discrete variables. The integer variable, as the name implies, is an integer value selected from the defined range bounds. Several engineering problems have both discrete and continuous design variables; such problems are said to be as mixed design variable problems.

1.3 CONSTRAINT HANDLING

When constraints occur in engineering problems, it is necessary to include supporting techniques within the methods. Several constraint handling techniques, such as penalty-based, probability-based and feasibility-based, have been proposed to date. As the choice of penalty parameter necessitates a significant number of preliminary trials, a parameter-less approach, referred to as a niched penalty function approach, was proposed by Deb and Agrawal (1999). In this approach, a feasible solution was selected based on three criteria: accepting the feasible solution rather than infeasible solution, accepting best fit solution from two feasible solutions and accepting infeasible solutions based on a smaller number of constraint violations. These three rules were then referred to as feasibility-based rules and were used as a constraint handling technique (Deb, 2000). The probability distribution-based constraint handling technique was proposed by Kulkarni and Shabir (2016) and Kulkarni et al. (2016) to ensure the fitness function was biased towards feasibility. However, the mathematical construction of this approach is problem dependent and needs to be generalised.

1.4 OVERVIEW OF COHORT INTELLIGENCE (CI) ALGORITHM

An AI-based optimisation technique, referred to as Cohort Intelligence (CI), was proposed by Kulkarni et al. (2013). It was motivated by the social behaviour of cohort candidates. The term 'cohort' refers to a group of learning candidates cooperating, competing and interacting with one another in order to achieve and improve their individual goal; this goal is inherently common to all the candidates. In the course of learning through interaction and competition, a candidate may follow some other candidate. This may result in the improvement of its own behaviour as well as that of the overall cohort. The cohort could be considered successful when the behaviour of every candidate is saturated and does not improve further.

The methodology of the CI algorithm incorporated with the static penalty function (SPF) approach is applied in solving several structural and engineering problems that have discrete, integer and mixed variables. These problems are classified into two domains: truss structure design problems with discrete variables and mechanical design engineering problems with mixed variables. The truss structure problems considered here are the 6-bar, 10-bar (2 cases), 25-bar (2 cases), 38-bar, 45-bar, 52-bar and 72-bar (2 cases) problems. The mechanical design engineering problems considered here are the reinforced concrete beam design problem, the steeped cantilever beam design problem, the welded beam problem (two cases), the speed reducer problem, the pressure vessel design problem, the helical compression spring problem, the multi-clutch brake problem, the I-section beam problem, the cantilever beam problem and the compound gear design problem. Furthermore, 17 linear and nonlinear integer variable test problems (linear, nonlinear, global, convex and monotonous functions) are also considered. The constraints involved in these problems are handled using a static penalty function (SPF) approach. A round off integer sampling approach is devised for handling the discrete variables. The results are compared with those obtained using the Multi Random Start Local Search (MRSLS) method proposed in Kulkarni et al. (2016). The performance of the CI algorithm is also evaluated for two parameters (the sampling space reduction factor (R) and the number of candidates (C)) using two illustrative examples (one from each problem domain) (see Section 3.3). To overcome the limitation of the SPF approach that the choice of penalty parameter (θ) necessitates a significant number of preliminary trials, as in Kale and Kulkarni (2018), the Self-Adaptive Penalty Function (SAPF) approach is proposed. The effect of the SAPF approach on outcomes such as the penalty function, constraint violations and the pseudo objective function are thoroughly discussed. Additionally, in order to overcome the limitations of the CI-SAPF approach (i.e., setting the sampling space reduction factor R) discussed in Section 3.3, it is hybridized with the CBO and referred to as the CI-SAPF-CBO. The proposed CI-SAPF and CI-SAPF-CBO are tested by solving 10 discrete truss structure problems, 11 mixed variable design engineering problems and 17 discrete variable test problems (linear, nonlinear, global, convex and monotonous functions). Previously (Kaveh and Mahdavi, 2014; Kale and Kulkarni, 2018), CBO was applied to solve certain truss structure and engineering design problems. In this book, CBO is also applied to solve other problems from similar domains and is considered here in order to compare the performance of individual algorithms. The CI-SPF, CI-SAPF, CI-SAPF-CBO and CBO algorithms were applied to two real-world applications. These problems were from the manufacturing engineering domain: (i) the multi-pass turning problem for minimisation of unit production cost and (ii) the multi-pass turning problem for the minimisation of production time. The solutions obtained from the proposed techniques are compared with other contemporary techniques discussed in the literature.

1.5 ORGANISATION OF THE BOOK

This book is organised as follows:

The detailed literature review of nature inspired optimisation techniques including the classification is given in Chapter 2. The review emphasises the classification of socio-based optimisation techniques and considers the pros and cons of various constraint handling methods.

In Chapter 3, the algorithms of CI and MRSLS are incorporated with SPF and are discussed in detail along with the solutions to discrete and mixed variable problems from the truss structure and design engineering domains. In order to handle discrete variables, a round off integer sampling approach is also proposed. The CI-SPF and MRSLS approaches are validated by solving the 6-bar truss structure problem, the pressure vessel design engineering problem and two linear and nonlinear test problems. The discussion of the results includes comparisons between these approaches and, additionally, those of other nature-inspired optimisation techniques. In addition, the CI algorithm is analysed by varying the number of candidates C and the sampling space reduction factor R and solving the discrete variable 10-bar truss structure problem and the mixed variable steeped beam design engineering problem.

In Chapter 4, the proposed SAPF approach is discussed in detail. The approach is validated by solving the problems discussed in Chapter 3. Using the SAPF approach, the impact of a penalty parameter on the behaviour of the function value, constraint violations and pseudo objective function are analysed.

As discussed in the literature review, the nature-inspired optimisation algorithms are governed by computational parameters. The algorithm for CI also requires two parameters to be tuned: the number of candidates C and the sampling space reduction factor R. The fine-tuning of such parameters necessitates several preliminary trials. In order to reduce the dependence on the sampling space reduction factor R, the intrinsic properties of CI and CBO are combined to form a new hybrid metaheuristic algorithm CI-SAPF-CBO (see Chapter 5). For the validation of the proposed CI-SAPF-CBO algorithm, one problem from each of the domains discussed in Chapter 3 is solved and the results compared to those obtained using other contemporary techniques.

In Chapter 6, the applicability of the CI-SPF, MRSLS, CI-SAPF and CI-SAPF-CBO techniques on different domain problems (9 discrete variable truss structure problems, 10 mixed variable design engineering problems and 15 discrete variable test problems (linear, nonlinear, global, convex and monotonous functions)) is compared to other contemporary techniques. The formulations of these problems are given in the Appendix.

In Chapter 7, the applicability of proposed the CI-SPF, CI-SAPF and CI-SAPF-CBO on real-world applications from the manufacturing engineering domain is discussed in detail. The problems on multi-pass turning and multi-pass milling process are solved. A comparison of the results with those obtained using other contemporary techniques is also presented.

REFERENCES

Colorni, A., Dorigo, M. and Maniezzo, V. (1991) 'Distributed optimization by ant colonies', *ECAL91 European Conference on Artificial Life*, Paris, pp. 134–142.

Deb, K. (2000) 'An efficient constraint handling method for genetic algorithms', *Computer Methods in Applied Mechanics in Engineering*, Vol. 186, Nos. 2–4, pp. 311–338.

Deb, K. and Agrawal, S. (1999) 'A niched-penalty approach for constraint handling in genetic algorithms', in *Proceedings of the International Conference on Artificial Neural Networks and Genetic Algorithms (ICANNGA-99)*, pp. 235–243.

Eberhart, R. and Kennedy, J. (1995) 'A new optimizer using particle swarm theory'. In: *MHS'95. Proceedings of the Sixth International Symposium on Micro Machine and Human Science*, IEEE, Nagoya, pp. 39–43.

Goldberg, D.E. (1989) 'Genetic algorithms, in *Search, Optimization and Machine Learning*, Addison-Wesley.

Kale, I.R. and Kulkarni, A.J. (2018) 'Cohort intelligence algorithm for discrete and mixed variable engineering problems', *International Journal of Parallel, Emergent and Distributed Systems*, Vol. 33, No. 6, pp. 627–662.

Kaveh, A. and Mahdavi V.R. (2014) 'Colliding bodies optimization: A novel meta-heuristic method', *Computers and Structures*, Vol. 39, 15 July, pp. 18–27.

Kulkarni, A.J. and Shabir H. (2016) 'Solving 0–1 knapsack problem using cohort intelligence algorithm', *International Journal of Machine Learning and Cybernetics*, Vol. 7, No. 3, pp. 427–441.

Kulkarni, A.J., Durugkar I.P. and Kumar M. (2013) 'Cohort intelligence: A self supervised learning behavior', *Systems, Man, and Cybernetics (SMC), IEEE International Conference*, pp. 1396–1400.

Kulkarni, A.J., Baki, M.F. and Chaouch, B.A. (2016) 'Application of the cohort-intelligence optimization method to three selected combinatorial optimization problems', *European Journal of Operational Research*, Vol. 250, No. 2, pp. 427–447.

Kumar, M., Kulkarni, A.J., Satapathy, S.C. (2018) 'Socio evolution and learning optimization algorithm: A socio-inspired optimization methodology', *Future Generation Computer Systems*, Vol. 81, pp. 252–272.

CHAPTER 2

Literature Survey on Nature Inspired Optimisation Methodologies and Constraint Handling

2.1 CLASSIFICATION OF NATURE INSPIRED OPTIMISATION TECHNIQUES

The classification of nature inspired optimisation techniques are presented in Figure 2.1 (taken from Kumar et al. 2018). The details are described as follows:

Bio-inspired Techniques: The bio-inspired metaheuristic Genetic Algorithm (GA) was proposed by John Holland in the 1960s (Holland, 1975) and was based on Darwin's theory of evolution, i.e., the survival of the fittest. The GA was further modified by David Goldberg and team (Goldberg, 1989). The algorithm relied on the three important biological operators of mutation, crossover and selection, which make the algorithm more able to approximate better quality solutions. An evolutionary intelligence is a set

DOI: 10.1201/9781003245193-2

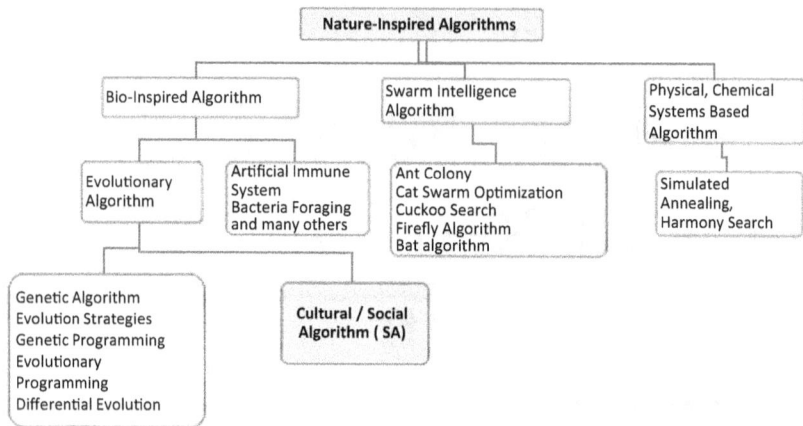

FIGURE 2.1   Classification of Nature-Inspired Optimisation Algorithms.

Source: Kumar et al., 2018.

of evolutionary strategies, such as evolutionary programming, differential evolution and neuro-evolution (Michalewicz et al., 1996). Such strategies are inspired by biological evolutionary strategies such as reproduction, mutation, recombination and selection. In evolutionary programming (Fogel et al., 1966), a stochastic finite element approach was used to evolve the generations, with mutation being the key operator utilised. Furthermore, a heuristic evolutionary strategy was proposed by Rechenberg (1971). This technique is dependent on both mutation and selection search operators. The mutation strength was governed by modifying the individual step size for each coordinate correlation using a covariance matrix adaptation approach.

Swarm Intelligence Techniques: Inspired by the flocking of birds and the behaviour of schools of fish, a stochastic Particle Swarm Optimisation (PSO) model was proposed by Eberhart and Kennedy (1995). Based on the intelligence exhibited by a swarm, several other techniques have also been proposed. These include Ant Colony Optimisation (ACO) (Colorni et al., 1991), the Firefly Algorithm (FA) (Yang, 2010a) and the Bat Algorithm (BA) (Yang, 2010b). The ACO algorithm models the foraging behaviour of ants, in which they search for a shortest path between food and their nest. This is possible due to the excretion of pheromones by the ants that enables them to follow every other ant in the system. In the FA, all the fireflies are unisex and attracted towards every other firefly based on the intensity of flash signals. Fireflies are attracted to areas of higher intensity (brightness) and move towards better search spaces by decreasing the distance between them. In a similar way to the FA, the behaviour of bats is based on the frequency of ultrasonic waves. The performance of these swarm algorithms is very similar to Multi Agent Systems (MAS) where all the agents work in a group to achieve the best possible outcome. These metaheuristics have been proven to be highly compatible in the solution of complex problems with both linear and nonlinear constraints (Gandomi et al., 2011; He and Wang, 2007; Kaveh and Talatahari, 2009).

Socio-inspired Techniques: The main socio-inspired optimisation techniques are the Probability Collective (PC) (Wolpert and Turner, 1999), Symbiosis Organism Search (SOS) (Cheng and Prayogo, 2014) and the Socio evolution and learning optimisation algorithm (SELO) (Kumar et al., 2018). The classification is illustrated in Figure 2.2 (taken from Kumar et al., 2018). The PC is a distributed and decentralised approach defined in the framework of Collective Intelligence (COIN). It decomposes the entire system into subsystems and treats them as a group of learning, rational and self-interested agents, or as a MAS. The SOS models symbiotic interaction

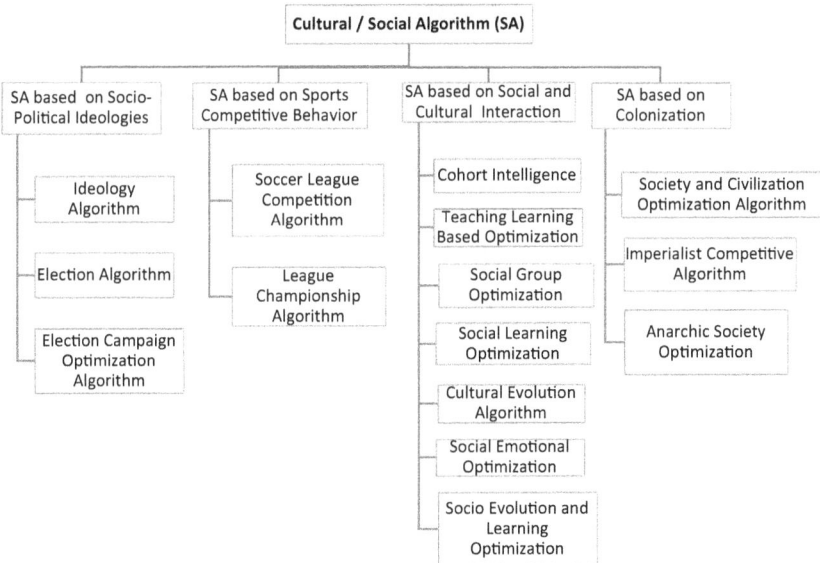

FIGURE 2.2   Classification of Socio-Inspired Algorithms.
Source: Kumar et al., 2018.

strategies that the independent agents (organisms) use to survive in the ecosystem. Several political election-based socio-algorithms such as the Ideology Algorithm (IA) (Teo et al., 2017), the Election Algorithm (EA) (Emami and Derakshan, 2015) and Election Campaign Optimisation (ECO) (Lv et al., 2010) have also been proposed. ECO models the social behaviour of voters when the candidate attempts to obtain maximum support from them. Based on the position of the candidate and voters, both global and local voters are considered. The uniform distribution process is used to identify the supported focus of the candidate. The EA is based on the process of candidate promotion during the election campaign. There is a series of steps that candidates use to work towards reinforcing their positive image and to compete with one another to increase their popularity. Additionally, candidates who have similar ideologies ally together to increase the chance of success of the united party. With similar motivation, the IA was proposed by Teo et al. (2017). This algorithm emphasises the behaviour of political parties aiming to improve their rank.

The League Championship Algorithm (LCA) (Kashan, 2009) is inspired by distinct features of a sporting activity and models the social tendencies of sport competition in a league. Using a similar approach to LCA, the Soccer League Competition (SLC) algorithm was proposed by Moosavian and Roodsari (2014). It is based on the interaction of players during a soccer match. The socio-inspired algorithms are also classified according to cultural interactions such as Teaching Learning Based Optimisation (TLBO) (Rao, 2011), which depicts the process of outcome-based education. The influence of teaching process on student outcome is modelled. The Social Group Optimisation (SGO) (Satapathy and Naik, 2016) and Social Learning Optimisation (SLO) (Liu et al., 2016) algorithms are based on the process of the propagation of human knowledge in the learning society/group to solve complex engineering problems. A Cultural Evolution Algorithm (CEO) (Kuo and Lin, 2013) is inspired by the evolution of social species. It adopts certain strategies such as group consensus, individual learning, innovative learning and self-improvement to evolve. The Social Emotional Optimisation (SEO) algorithm (Xu et al., 2010) is another swarm-based socio-inspired metaheuristic which simulates an individual who wishes to achieve a higher status in the society and whose decisions are guided by their emotions. The level of emotion is based on the index (supporting parameter) that controls its current behaviour, and the society decides whether its current behaviour is better or worse and the emotion index value is affected accordingly. SELO is the social learning of humans organised as families in a society. It models the social evolution and learning of parents and children who constitute a family. Individuals organised as family groups (parents and children) interact with one another and other distinct families to attain predefined individual goals.

The Society and Civilization Optimisation (SCO) Algorithm (Ray and Liew, 2003) is inspired by human social behaviour seen among society individuals. The individuals in a society interact with one another to improve their overall behaviour and a cooperative interaction among such societies represents a civilization. The Imperialistic Competitive Algorithm (ICA) (Atashpaz-Gargari and Lucas, 2007) simulates socio-political behaviours seen across imperialist nations which compete to take possession of weaker colonies or empires. This imperialist competition results in the power of stronger and successful imperialist empires being enhanced, whilst weaker empires gradually collapse, leading to a state of convergence. Another optimisation algorithm which seeks inspiration from the commonly observed human behaviours where greed and disorder are used to achieve goals is referred to as the Anarchic Society Optimisation (ASO) algorithm (Ahmadi-Javid, 2011).

2.2 BACKGROUND OF THE COHORT INTELLIGENCE ALGORITHM

Using a similar approach to the socio-inspired techniques discussed earlier, the Cohort Intelligence (CI) algorithm was proposed by Kulkarni et al. (2013). It is motivated by the social learning behaviour of candidates such as following, interacting, cooperating and competing with every other candidate in the cohort. Initially, the CI algorithm was tested on unconstrained benchmark test examples (Kulkarni et al., 2013). It was then implemented for constrained problems and applied to solve the combinatorial NP-hard 0-1 Knapsack problem with the number of items varying from 4 to 75 (Kulkarni and Shabir, 2016). The constraints involved in this problem were handled by a problem-specific probability-based constraint handling technique. The algorithm yielded competent results as compared to integer programming solutions. This approach was also applied when solving real-world combinatorial problems from the healthcare and logistics domain as well as for large-sized complex problems from the Cross Border Supply Chain domain (Kulkarni et al.,

2016a), the Traveling Salesman Problem (TSP) (Kulkarni et al., 2017) and several other benchmark problems (Kulkarni et al., 2017). The algorithm performed significantly better than both integer programming and other specific heuristics techniques. Krishnasamy et al. (2014) modified the CI algorithm by incorporating a mutation mechanism, which helped in expanding the sampling space by introducing diversity and, additionally, avoided premature convergence. The modified CI was compared with the original CI for solving several clustering problems. In addition, it was hybridized with a K-Means algorithm which also exhibited superior performance. Gaikwad et al. (2015) proposed a Modified Analytical Hierarchy Process (MAHP) which was combined with GA and CI to identify the level of sugar in ice-cream for diabetic patients. The constraint handling for the static and dynamic penalty function approaches were incorporated in the CI (CI-SPF and CI-DPF, respectively) for solving several test problems and manufacturing engineering problems (Kulkarni et al., 2016c). The CI-SPF was adopted for solving complex problems from truss structure and mechanical engineering domain (Kale and Kulkarni, 2018). Furthermore, Patankar and Kulkarni (2017) introduced seven variations of the CI algorithm. The variations were associated with the choices candidates made when selecting other candidates from which to learn certain characteristics. They are labelled as: follow best, follow better, follow worst, follow itself, follow median, follow roulette wheel selection, and alienation and random selection. The algorithm was tested on several unimodal and multimodal unconstrained problems. Moreover, CI was also applied to the security of secret messages using steganography (Sarmah and Kulkarni, 2017, 2018). The three cases of shell-and-tube heat exchanger design problem were also solved for minimisation of cost and obtained significantly better results as compared to other contemporary algorithms (Dhavle et al., 2016).

As with other nature inspired methods, the performance of CI deteriorates when constraints are involved. To date, a dynamic penalty function (DPF) (Kulkarni et al., 2016b) has been proposed; however, the choice of penalty parameter requires a significant number of preliminary trials of the algorithm. Also, a probability-based constraint handling approach was proposed by Kulkarni and Shabir (2016). The approach is problem specific and may become tedious when the number of constraints is increased. The review of various constraint handling approaches is thoroughly discussed in the next section.

2.3 LITERATURE REVIEW ON CONSTRAINT HANDLING TECHNIQUES

The algorithms discussed in Section 2.1 were incorporated with constraint handling techniques such as penalty function methods, probability distribution and feasibility-based rule. The penalty function techniques are employed to handle constraints when using a certain penalty value and convert the constrained problem into an unconstrained problem. As the choice of suitable penalty parameter necessitates a significant number of preliminary trials, a parameter-less approach, referred to as a niched penalty function approach, was proposed by Deb and Agrawal (1999). In this approach, a feasible solution was selected based on three criteria: accept the feasible solution rather than infeasible solution; accept the best fit solution from two feasible solutions; and accept an infeasible solution based on fewer constraint violations. These three rules were then referred to as the feasibility-based rule and were used as a constraint handling technique (Deb, 2000). The probability distribution-based constraint handling technique was proposed to make fitness function bias towards feasibility (Kulkarni and Shabir, 2016; Kulkarni et al., 2016b). However, the mathematical construction of this approach is problem dependent and needs to be generalised.

The penalty function approach has been widely used due to its simple construction and ease of implementation. Several penalty-based constraint handling techniques have been proposed so far, such as the barrier (death) penalty function approach, which was based on the elimination of infeasible solutions (Luenberger and Ye, 2016), the exact penalty function (Homaifar et al., 1994) and the dynamic penalty function (Joines and Houck, 1994) based on setting the value of the penalty parameter and its multiplication factors (penalty reduction or expansion factor) to penalise the objective function. Other techniques were also proposed such as the annealing penalty function (Michalewicz and Attia, 1994; Carlson and Shonkwiler, 1998), which was based on idea of Simulated Annealing (SA), and the adaptive penalty function (Gen and Cheng, 1996; Hadj-Alouane and Bean, 1997; Smith and Tate, 1993; Yokota et al., 1996), which was aimed at eliminating the setting of penalty parameter from other penalty function approaches. In penalty-based segregated GA (Le Riche et al., 1995), a distinct penalty parameter was set for different evaluated fitness functions. These techniques were successfully employed with nature inspired optimisation techniques in order to deal with linear and nonlinear constrained optimisation problems. These techniques are simple and easy to apply to a wide variety of constrained optimisation problems (Yu et al., 2010; Li et al., 2011); however, as the number of constraints increase, their performance degenerates (Luenberger and Ye, 2016). Additionally, an exact penalty approach was adopted by Shin et al. (1990), Wu and Chow (1995) and Azad et al. (2013) for nonlinear optimisation problems with discrete design variables. For every independent problem, several preliminary trials were required to set an appropriate penalty parameter (Homaifar et al., 1994; Morales and Quezada, 1998). A similar approach was adopted for FA (Gandomi et al., 2011) and the CI algorithm with static penalty function approach (CI-SPF) (Kale and Kulkarni, 2018) for solving discrete and mixed variable problems with linear as well as nonlinear constraints from engineering design and truss structure domains. However, it was noticed that the selection of penalty parameter becomes tedious with the increase in number of constraints.

Kannan and Kramer (1994) adopted an augmented Lagrange Multiplier approach (of Viswanathan and Grossmann, 1990) incorporated with dynamic penalty function method for solving discrete and mixed variable problems from the design engineering domain. In this approach, the penalty parameter was multiplied by a suitable factor and iteratively penalised the objective function. A generalised Hopfield network using an extended penalty function approach was proposed by Shih and Yang (2002) to solve nonlinear engineering problems with discrete and mixed variables. In these methods, the penalty parameter was initialised based on arbitrary value (0 or 1) and then updated iteratively with an incremental multiplication factor. A limitation was observed that if the multiplication factor is too high, then the objective function value may become unstable and stuck at the local minima. In a similar way to the dynamic penalty function approach, Curtis and Nocedal (2008) introduced a flexible penalty function to handle nonlinear constraints. Here, the penalty parameter was arbitrarily chosen from the prescribed interval rather than choosing a fixed value which influentially guided the convergence.

Nanakorn and Meesomklin (2001) proposed an adaptive penalty function approach in which the modified binary scaling technique was employed to scale the fitness value. This fitness value then scaled in three different categories: the minimum fitness value, the average fitness of all feasible value and the best feasible value. Every infeasible value was then penalised by the best infeasible value having scaled fitness equal to φ times that of average fitness value. The parameter φ needed to be set based on preliminary trials. Broyden and Attia (1984) proposed a smooth sequential penalty function incorporated with a Quasi Newton approach and then it was combined with orthogonal transformations based on Jacobian constraints. The non-stationary multi-stage penalty function approach was implemented by Parsopoulos and Vrahatis (2002) and then followed by Coath and Halgamuge (2003) with a feasibility preservation method for solving nonlinear problems.

Using an evolutionary algorithm, Michalewicz et al. (1996) and Coello (2000) proposed an approach in which the penalty function is split into two distinct parts such as sum of violations the constraints have and the number of constraints that are violated and then penalty was individually applied. For this approach, independent weighting factors needed to be chosen that increased the number of working parameters. In addition, several initial trials were required to be set up using suitable penalty parameters for both parts. Nie (2006) proposed a new approach of semi-penalty that considered the qualities of the Sequential Quadratic Programming (SQP) method and the Sequential Penalty Quadratic Programming (SPQP) method where equality and inequality constraints received distinct treatment.

An external penalty function scheme was adopted by Hasançebi and Azad (2015) in different manner, where a relaxation strategy was incorporated into the Adaptive Dimensional Search (ADS) method. In this strategy, the infeasible solution was penalised and dominated by selecting an infeasible solution in order to escape from the local minima. Once the solution is saturated, the intensity of the penalty parameter was reduced by multiplying the reduction factor. After every Stagnation Escape Period (SEP), the solution was recalculated using an updated penalty parameter and then compared with the previous saturated solution. It was observed that, if the recalculated solution was worse than the previous solution, the original penalty parameter was utilised to recalculate the solution and continue the process in the search of best solution. In order to update the penalty parameter, an additional multiplication factor need to be included. This may require additional time to set the multiplication factor, which tends to increase the overall computational cost. Furthermore, the feasibility-based constraint handling approach proposed in Deb (2000) was later implemented by Bansal et al. (2009), Kaveh and Talatahari (2009) and Kulkarni and Tai (2011). It was further modified by Kulkarni et al. (2016a) where a worse solution after a stagnation period was successfully accepted, which helped the algorithm jump out of the local minima. It was success-fully applied to solving problems from design engineering and truss structure domain.

2.4 CONCLUSION

This chapter provides a detailed literature survey on nature inspired optimisation techniques and constrained handling techniques. The classification of nature inspired optimisation techniques is presented. The socio-inspired approach is one of the subdomains of nature inspired optimisation techniques; hence, its classification is also presented and the literature discussed in detail. In this book, the constrained problems various domains are considered in order to validate the proposed techniques (presented in Chapter 3, 4 and 5); the merits and demerits of various constraint handling techniques are also discussed.

REFERENCES

Ahmadi-Javid, A. (2011) 'Anarchic society optimization: A human-inspired method', *Evolutionary Computation (CEC), 2011 IEEE Congress*, New Orleans, pp. 2586–2592.

Atashpaz-Gargari, E., and Lucas, C. (2007) 'Imperialist competitive algorithm: An algorithm for optimization inspired by imperialistic competition', *Evolutionary Computation (CEC), 2007 IEEE Congress*, Singapore, pp. 4661–4667.

Azad, S.K., Hasançebi, O., Azad, S.K. and Erol, O.K. (2013) 'Upper bound strategy in optimum design of truss structures: A big bang-big crunch algorithm based application', *Advances in Structural Engineering*, Vol. 16, No. 6, pp. 1035–1046.

Bansal, S., Mani, A. and Patvardhan, C. (2009) 'Is Stochastic ranking really better than feasibility rules for constraint handling in evolutionary algorithms?', *Proceedings of World Congress on Nature and Biologically Inspired Computing*, pp. 1564–1567.

Broyden, C.G. and Attia N.F. (1984) *A Smooth Sequential Penalty Function Method for Solving Nonlinear Programming Problems*, Springer.

Carlson, S.E. and Shonkwiler, R. (1998) 'Annealing a genetic algorithm over constraints', *SMC'98 Conference Proceedings, IEEE International Conference on Systems, Man, and Cybernetics (Cat. No. 98CH36218)*, Vol. 4, pp. 3931–3936.

Cheng, M.Y. and Prayogo, D. (2014) 'Symbiotic organisms search: a new metaheuristic optimization algorithm', *Computers and Structures*, Vol. 139, pp. 98–112.

Coath, G. and Halgamuge, S.K. (2003) 'A comparison of constraint-handling methods for the application of particle swarm optimization to constrained nonlinear optimization problems', *Evolutionary Computation*, Vol. 4, pp. 2419–2425.

Coello, C.A.C. (2000) 'Use of a self-adaptive penalty approach for engineering optimization problems', *Computers in Industry*, Vol. 41, pp. 113–127.

Colorni, A., Dorigo, M. and Maniezzo, V. (1991) 'Distributed optimization by ant colonies', *ECAL91 European Conference on Artificial Life*, Paris, pp. 134–142.

Curtis, F.E. and Nocedal, J. (2008) 'Flexible penalty functions for nonlinear constrained optimization', *IMA Journal of Numerical Analysis*, Vol. 28, No. 4, pp. 749–769.

Deb, K. (2000) 'An efficient constraint handling method for genetic algorithms', *Computer Methods in Applied Mechanics in Engineering*, Vol. 186, Nos. 2–4, pp. 311–338.

Deb, K. and Agrawal, S. (1999) 'A niched-penalty approach for constraint handling in genetic algorithms', *Proceedings of the International Conference on Artificial Neural Networks and Genetic Algorithms (ICANNGA-99)*, pp. 235–243.

Dhavle, S.V., Kulkarni, A.J., Shastri, A. and Kale, I.R. (2016) 'Design and economic optimization of shell-and-tube heat exchanger using cohort intelligence algorithm', *Neural Computing and Applications*, Vol. 30, No. 1, pp. 111–125.

Eberhart, R. and Kennedy, J. (1995) 'A new optimizer using particle swarm theory', in *MHS'95. Proceedings of the Sixth International Symposium on micro Machine and Human Science, IEEE*, Nagoya, pp. 39–43.

Emami, H. and Derakhshan, F. (2015) 'Election algorithm: A new socio-politically inspired strategy', *AI Communications*, Vol. 28, No. 3, pp. 591–603.

Fogel, L.J., Owens, A.J., Walsh, M.J. (1966) *Artificial Intelligence through Simulated Evolution*, John Wiley.

Gaikwad, S.M. , Joshi, R.R. and Kulkarni, A.J. (2015) 'Cohort intelligence and genetic algorithm along with AHP to recommend an ice cream to a diabetic patient', *International Conference on Swarm, Evolutionary, and Memetic Computing*, pp. 40–49.

Gandomi, A.H., Yang, X-S. and Alavi, A.H. (2011) 'Mixed variable structural optimization using firefly algorithm', *Computers and Structures*, Vol. 89, Nos. 23–24, pp. 2325–2336.

Gen, M. and Cheng, R. (1996) 'A survey of penalty techniques in genetic algorithms', *Proceedings of International Conference on Evolutionary Computation, IEEE*, pp. 804–809.

Goldberg, D.E. (1989) *Genetic Algorithms in Search, Optimization and Machine Learning*, Addison-Wesley.

Hadj-Alouane, A.B. and Bean, J.C. (1997) 'A genetic algorithm for the multiple-choice integer program', *Operations Research*, Vol. 45, pp. 92–101.

Hasançebi, O. and Azad, S.K. (2015) 'Adaptive dimensional search: A new metaheuristic algorithm for discrete truss sizing optimization', *Computers and Structures*, Vol. 154, pp. 1–16.

He, Q. and Wang, L. (2007) 'An effective co-evolutionary particle swarm optimization for constrained engineering design problem', *Engineering Applications of Artificial Intelligence*, Vo1. 20, No. 1, pp. 89–99.

Holland, J.H. (1975) *Adaptation in Natural and Artificial Systems: An Introductory Analysis with Applications to Biology, Control, and Artificial Intelligence*, University Michigan Press.

Homaifar, A., Lai, S.H.Y. and Qi, X. (1994) 'Constrained optimization via genetic algorithms', *Simulation*, Vol. 62, No. 4, pp. 242–254.

Joines, J. and Houck, C. (1994) 'On the use of non-stationary penalty functions to solve non-linear constrained optimization problems with gas', *Proceedings of the First IEEE International Conference on Evolutionary Computation*, pp. 579–584.

Kale, I.R. and Kulkarni, A.J. (2018) 'Cohort intelligence algorithm for discrete and mixed variable engineering problems', *International Journal of Parallel, Emergent and Distributed Systems*, Vol. 33, No. 6, pp. 627–662.

Kannan, B.K. and Kramer, S.N. (1994) 'An augmented Lagrange multiplier based method for mixed integer discrete continuous optimization and its applications to mechanical design', *Journal of Mechanical Design*, Vol. 116, No. 2, pp. 405–411.

Kashan, A.H. (2009) 'League championship algorithm: A new algorithm for numerical function optimization', *IEEE International Conference of Soft Computing and Pattern Recognition, Malacca, Malaysia*, pp. 43–48.

Kaveh, A. and Talatahari, S. (2009) 'A particle swarm ant colony optimization for truss structures with discrete variables', *Journal of Constructional Steel Research*, Vol. 65, pp. 1558–1568.

Krishnasamy, G., Kulkarni A.J. and Paramesran, R. (2014) 'A hybrid approach for data clustering based on modified cohort intelligence and K-means', *Expert Systems with Applications*, Vol. 41, pp. 6009–6016.

Kulkarni, A.J. and Shabir, H. (2016) 'Solving 0–1 knapsack problem using cohort intelligence algorithm', *International Journal of Machine Learning and Cybernetics*, Vol. 7, No. 3, pp. 427–441.

Kulkarni, A.J. and Tai, K. (2011) 'A probability collectives approach with a feasibility-based rule for constrained optimization', *Applied Computational Intelligence and Soft Computing*, Article ID 980216.

Kulkarni, A.J., Durugkar, I.P. and Kumar, M. (2013) 'Cohort intelligence: A self supervised learning behavior', *Systems, Man, and Cybernetics (SMC), IEEE International Conference*, pp. 1396–1400.

Kulkarni, A.J., Kale, I.R. and Tai, K. (2016a) 'Probability collectives for solving discrete and mixed variable problems', *International Journal of Computer Aided Engineering and Technology*, Vol. 8 No. 4, pp. 325–361.

Kulkarni, A.J., Baki, M.F. and Chaouch, B.A. (2016b) 'Application of the cohort-intelligence optimization method to three selected combinatorial optimization problems', *European Journal of Operational Research*, Vol. 250, No. 2, pp. 427–447.

Kulkarni, O., Kulkarni, N., Kulkarni, A.J. and Kakandikar, G. (2016c) 'Constrained cohort intelligence using static and dynamic penalty function approach for mechanical components design', *International Journal of Parallel, Emergent and Distributed Systems*, Vol. 33, No. 6, pp. 1–19.

Kulkarni, A.J., Krishnasamy, G. and Abraham, A. (2017) *Cohort Intelligence: A Socio-Inspired Optimization Method*, Springer.

Kumar, M., Kulkarni, A.J., Satapathy, S.C. (2018) 'Socio evolution and learning optimization algorithm: A socio-inspired optimization methodology', *Future Generation Computer Systems*, Vol. 81, pp. 252–272.

Kuo, H.C. and Lin, C.H. (2013) 'Cultural evolution algorithm for global optimizations and its applications', *Journal of Applied Research and Technology*, Vol. 11, No. 4, pp. 510–522.

Le Riche, R., Knopf-Lenoir, C. and Haftka, R.T. (1995) 'A segregated genetic algorithm for constrained structural optimization', *Proceedings of the Sixth International Conference on Genetic Algorithms*, pp. 558–565.

Li, B., Yu, C.J., Teo, K.L. and Duan, G.R. (2011) 'An exact penalty function method for continuous inequality constrained optimal control problem', *Journal of Optimal Theory and Applications, Springer*, Vol. 151, pp. 260–291.

Liu, Z.Z., Chu, D.H., Song, C., Xue, X. and Lu, B.Y. (2016) 'Social learning optimization (SLO) algorithm paradigm and its application in QoS-aware cloud service composition', *Information Sciences*, Vol. 326, pp. 315–333.

Luenberger, D.G., and Ye, Y. (2016) 'Penalty and barrier methods', in *Linear and Nonlinear Programming*, Springer, Vol. 228.

Lv, W., Liu, Z., Zhang, X., Luo, S. and Cheng, S. (2010) 'Election campaign algorithm', *2nd International Asia Conference on Informatics in Control, Automation and Robotics*, Vol. 2, pp. 71–74.

Michalewicz, Z. and Attia, N. (1994) 'Evolutionary optimization of constrained problems', *Proceedings of the Third Annual Conference on Evolutionary Programming, World Scientific*, pp. 98–108.

Michalewicz, Z., Dasgupta, D., Le Riche, R. and Schoenauer, M. (1996) 'Evolutionary algorithms for constrained engineering problems', *Computers and Industrial Engineering Journal*, Vol. 30, No. 4, pp. 851–870.

Moosavian, N. and Roodsari, B.K. (2014) 'Soccer league competition algorithm: A novel meta-heuristic algorithm for optimal design of water distribution networks', *Swarm and Evolutionary Computation*, Vol. 17, pp. 14–24.

Morales, A.K. and Quezada, C.V. (1998) 'A universal eclectic genetic algorithm for constrained optimization', In *Proceedings of the 6th European Congress on Intelligent Techniques and Soft Computing*, Vol. 1, pp. 518–522.

Nanakorn, P. and Meesomklin, K. (2001) 'An adaptive penalty function in genetic algorithms for structural design optimization', *Computer and Structures*, Vol. 79, pp. 2527–2539.

Nie, P.Y. (2006) 'A new penalty method for nonlinear programming', *Computers and Mathematics with Applications*, Vol. 52, pp. 883–896.

Parsopoulos, K. and Vrahatis, M. (2002) 'Particle swarm optimization method for constrained optimization problems', *Intelligent Technologies Theory and Applications: New Trends in Intelligent Technologies*, Vol. 76, No. 1, pp. 214–220.

Patankar, N.S., Kulkarni, A.J. (2017) 'Variations of cohort intelligence', *Soft Computing*, Vol. 22, No. 6, pp. 1731–1747.

Rao, R.V. (2011) *Advance Modeling and Optimization of Manufacturing Processes*, Springer.

Ray, T. and Liew, K.M. (2003) 'Society and civilization: An optimization algorithm based on the simulation of social behavior', *IEEE Transactions on Evolutionary Computation*, Vol. 7, No. 4, pp. 386–396.

Rechenberg, I. (1971) 'Evolutions strategie – Optimierung technischer Systeme nach Prinzipien der biologischen Evolution', PhD thesis. Reprinted by Frommann-Holzboog (1973).

Sarmah, D.K. and Kulkarni, A.J. (2017) 'Image steganography capacity improvement using cohort intelligence and modified multi-random start local search methods', *Arabian Journal for Science and Engineering*, pp. 1–24.

Sarmah, D.K. and Kulkarni, A.J. (2018) 'JPEG based steganography methods using cohort intelligence with cognitive computing and modified multi random start local search optimization algorithms', *Information Sciences*, Vol. 430, pp. 378–396.

Satapathy, S. and Naik, A. (2016) 'Social group optimization (SGO): A new population evolutionary optimization technique', *Complex and Intelligence Systems*, Vol. 2, pp. 173–203.

Shih, C.J. and Yang, Y.C. (2002) 'Generalized Hopfield network based structural optimization using sequential unconstrained minimization technique with additional penalty strategy', *Advances in Engineering Software*, Vol. 33, No. 7-10, pp. 721–729.

Shin, D.K., Gurdal, Z. and Griffin, O.H. (1990) 'A penalty approach for nonlinear optimization with discrete design variables', *Engineering Optimization*, Vol. 16, No. 1, pp. 29–42.

Smith, A. and Tate, D. (1993) 'Genetic optimization using a penalty function', *Proceedings of the Fifth International Conference on Genetic Algorithms*, Morgan Kaufmann, pp. 499–503.

Teo, T.H., Kulkarni, A.J., Kanesan, J., Chuah, J.H. and Abraham, A. (2017) 'Ideology algorithm: A socio-inspired optimization methodology', *Neural Computing and Applications*, Vol. 28, No. 1, pp. 845–876.

Viswanathan, J. and Grossmann, I.E. (1990) 'A combined penalty function and outer-approximation method for MINLP optimization', *Computers and Chemical Engineering*, Vol. 14, No. 7, pp. 769–782.

Wolpert, D.H. and Tumer, K. (1999) 'An introduction to collective intelligence', Technical Report, NASA ARC-IC-99-63, NASA Ames Research Center.

Wu, S.J. and Chow, P.T. (1995) 'Steady-state genetic algorithms for discrete optimization of trusses', *Computers and Structures*, Vol. 56, No. 6, pp. 979–991.

Xu, Y., Cui, Z. and Zeng, J. (2010) 'Social emotional optimization algorithm for nonlinear constrained optimization problems', *Swarm, Evolutionary, and Memetic Computing (SEMCCO 2010), Lecture Notes in Computer Science, Springer Berlin Heidelberg*, Vol. 6466, pp. 583–590.

Yang, X.S. (2010a) 'Firefly algorithm, stochastic test functions and design optimisation', *International Journal of Bio-inspired Computation*, Vol. 2, No. 2, pp. 78–84.

Yang, X.S. (2010b) 'A new metaheuristic bat-inspired algorithm', *Nature Inspired Cooperative Strategies for Optimization (NICSO 2010)*, pp. 65–74.

Yokota, T., Gen, M., Ida, K., and Taguchi, T. (1996) 'Optimal design of system reliability by an improved genetic algorithm', *Electronics and Communications in Japan (Part III: Fundamental Electronic Science)*, Vol. 79, No. 2, pp. 41–51.

Yu, C., Teo, K.L., Zhang, L. and Bai, Y. (2010) 'A new exact penalty function method for continuous inequality constrained optimization problems', *Journal of Industrial and Management Optimization*, Vol. 6, No. 4, pp. 895–910.

CHAPTER 3

Cohort Intelligence (CI) Using the Static Penalty Function (SPF) Approach

3.1 CI-SPF

The CI method is inspired by the social tendency of cohort candidates to learn through interaction and competition with every other candidate. Every candidate may follow a certain behaviour within the cohort, which may result in improvement in its own behaviour. Certain beneficial qualities result in a particular behaviour which, when a candidate follows the behaviour, the candidate tries to adopt the associated qualities. This results in candidates learning from one another and helps the overall cohort behaviour to evolve. The cohort behaviour could be considered saturated/converged if, for a considerable number of learning attempts, the individual behaviour of each candidate does not improve considerably and also becomes almost same.

3.1.1 Framework of CI

In the context of one CI candidate following another candidate who is probabilistically chosen from the cohort using a roulette wheel approach. The CI algorithm (Kulkarni et al., 2013) is mathematically expressed as follows:

DOI: 10.1201/9781003245193-3

Step 1: Consider a cohort with C candidates; every individual candidate $c(c = 1, 2, ..., C)$ contains a set of attributes/variables $X^c = (X_1^c, X_2^c, ..., X_i^c ..., X_n^c)$ which makes the behaviour of an individual candidate $f(X^c)$. The initial solution is randomly generated in a way similar to the other population-based technique as follows:

$$X^c = \Psi^{lower} + \left(\Psi^{upper} - \Psi^{lower}\right).rand(1, n) \tag{3.1}$$

Step 2: A SPF approach is incorporated in order to handle the constraints and obtain the pseudo-objective function $\phi(X^c)$ (see Section 3.1.2).

Step 3: The probability of selecting behaviour $f(X^c)$ of every associated candidate $c(c = 1, 2, ..., C)$ is calculated as follows:

$$p^c = \frac{1 \Big/ \phi(X^c)}{\sum_{c=1}^{C} 1 \Big/ \phi(X^c)} \tag{3.2}$$

Step 4: Every individual candidate $c(c = 1, 2, ..., C)$ generates a random number $r \epsilon [0, 1]$ and using the roulette wheel approach decides to follow the corresponding behaviour and associated attributes.

Step 5: Every candidate $c(c = 1, 2, ..., C)$ shrinks the sampling interval Ψ_i^c, $i = 1, ..., n$ associated with every variable X_i^c, $i = 1, ..., n$ to its local neighbourhood. This is done as follows:

$$\Psi_i^c \epsilon \left[X_i^c - \left(\|\Psi_i\| / 2\right), X_i^c + \left(\|\Psi_i\| / 2\right)\right]$$

where $\Psi_i = \|\Psi_i\| \times R$ and R is the sampling space reduction factor

Each candidate $c(c = 1, 2, ..., C)$ samples their qualities from within the updated sampling interval Ψ_i^c and computes the function values. This makes the cohort available with C updated behaviours represented as $F^C = \left\{ f(X^1), ..., f(X^c), ..., f(X^C)\right\}$.

Step 6: The cohort behaviour can be considered saturated if there is no significant improvement in the behaviour $f(X^c)$ of every candidate. If either of the two criteria listed below is valid, accept any of the C behaviours from current set of behaviours in the cohort as the final objective function values as the final solution and stop, else return to Step 1.

(a) If the maximum number of attempts is exceeded.

(b) The cohort reaches a saturation state. There is no significant improvement in the further learning attempts.

3.1.2 Static Penalty Function (SPF) Approach

In general, the constrained optimisation problem is expressed as follows:

$$\text{Minimise } f(X) = f(X_1, X_2, \ldots, X_i, \ldots, X_n) \tag{3.3}$$

subject to

$$g_i(X) \leq 0, \quad i = 1, 2, \ldots, p$$

$$h_i(X) = 0, \quad i = 1, 2, \ldots, m$$

$$\Psi_i^{lower} \leq X_i \leq \Psi_i^{upper}$$

An exterior SPF constraint handling approach has been widely used (Goldberg, 1989). It is expressed as follows:

$$PF = \theta \times \left(\sum_{i=1}^{p} g_i(X) + \sum_{i=1}^{m} h_i(X) \right) \tag{3.4}$$

where θ is a penalty parameter and $\left(\sum_{i=1}^{p} g_i(X) + \sum_{i=1}^{m} h_i(X) \right)$ is the summation of the violated constraints. The value of θ needs to be chosen arbitrarily. The pseudo code for CI-SPF is illustrated in Figure 3.1.

3.1.3 A Round Off Integer Sampling Approach

In this work, the problems considered are from the truss structure and design engineering domains and have discrete and mixed design variables. In a similar way to that in Li et al. (2009), a round off integer sampling approach is incorporated in order to deal with discrete variables. The round off integer sampling approach works on the arbitrarily generated

C	Number of candidates ($c = 1,\dots,C$)
t	Number of variations
Ψ_i	Sampling space ($i = 1,\dots,n$)
R	Sampling space reduction factor
θ	Penalty parameter
P	Penalty function
g	Constraint value

Initialise C, t, Ψ_i, R, θ

While

1. Evaluate individual behaviour/objective function $f(\mathbf{X}^c)$
2. Apply penalty function approach to generate a pseudo behaviour/objective function:

$$\phi(\mathbf{X}^c) = f(\mathbf{X}^c) + P(\mathbf{X}^c)$$

where $P(\mathbf{X}^c) = \theta \times \sum_{c=1}^{C} g(\mathbf{X}^c)$

3. The probability p^c associated with every candidate c in the cohort is calculated as:

$$p^c = \frac{1 / \phi^*(\mathbf{X}^c)}{\sum_{c=1}^{C} 1 / \phi^*(\mathbf{X}^c)}$$

4. Using a roulette wheel approach, every candidate c selects the behaviour to follow from C available choices.
5. Every candidate c shrinks the sampling interval Ψ_i of every quality i in its neighbourhood.
6. **If:** No significant improvement in the behaviour $\phi^*(\mathbf{X}^c)$ is observed the solution is considered to be saturated

   Every candidate c expands the sampling interval Ψ_i associated with every quality i to its original interval Ψ_i

   Accept the current behaviour $\phi^*(\mathbf{X}^c)$ and the associated qualities.

   Else

   Return to Step 3

   End If

   End While

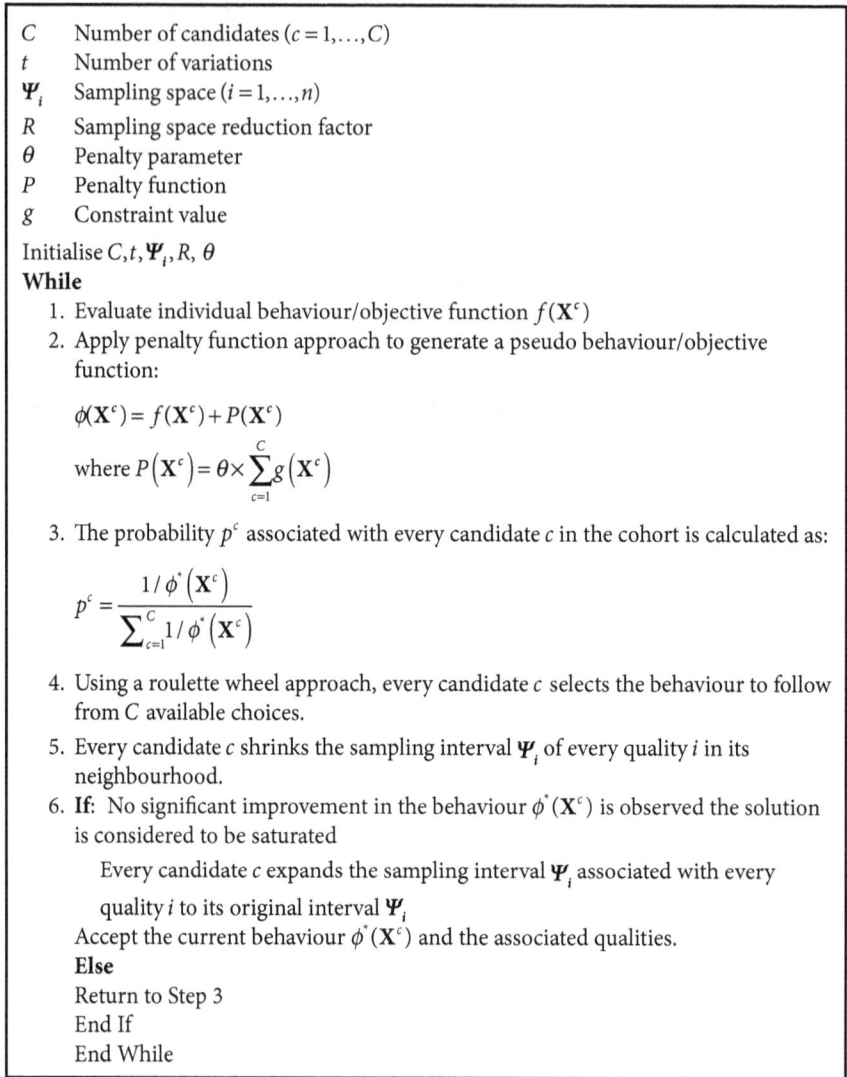

FIGURE 3.1   CI-SPF Pseudo Code.

integers' indices within the count of set of discrete variables. Based on these indices, actual variable values are selected from the given set. The details are discussed in following steps:

Step 1: Given a set of discrete values, $D = \{d1, d2, \dots, dm\}$; from which the variable vector, $X = \{X1, X2, \dots, Xn\}$ ($n <= m$) is to be initialized.

Step 2: Using uniform sampling, generate a set of n random values within 1 and m.

Step 2: Apply uniform sampling method to initialize five random numbers between 1 and 10	1.4456	4.5633	4.3822	2.3045	2.6563
Step 3: Round off to the nearest integer R	1	5	4	2	3
Step 4: Associated values within the original sampling space are chosen X	$X_1 = 0.5$	$X_2 = 2.5$	$X_3 = 3$	$X_4 = 1$	$X_5 = 1.5$

FIGURE 3.2   Illustration of the Round off Integer Sampling Approach.

Step 3: Round off the above set to the set of nearest integers, $R = \{r_1, r_2,..., r_n\}$.
Step 4: Finally, initialize the design variables as $X_j = d_{rj}$; $j = 1, 2, ..., n$.

This method is illustrated in Figure 3.2. Consider $D = \{0.5, 1, 1.5, 2, 2.5, 3, 3.5, 4, 4.5, 5\}$ having ten discrete values; within which suppose five variables $X = \{X_1, X_2, X_3, X_4, X_5\}$ need to be initialized.

3.2 TEST EXAMPLES

In this book, the CI-SPF algorithm with round-off integer sampling approach is successfully applied for solving mixed variable mechanical design engineering and discrete variable truss structural optimisation problems (see the Appendix). Here, one problem from each domain is presented; for example, the 6-bar truss structure problem from the structural domain, the pressure vessel design problem from the design engineering domain, the linear integer programming problem and the Rosen–Suzuki from the linear and nonlinear problem domain. In addition, 9 discrete variable truss structure problems, 10 mixed variable engineering design problems and 15 linear and nonlinear benchmark problems (linear, nonlinear, global, convex and monotonous functions) have also been solved (see Chapter 6). These problems are well studied in the literature and used to compare the performance of various optimisation algorithms such as GA (Deb et al., 1996; Wu and Chow, 1995; Rajeev and Krishnamoorthy, 1992; Yun, 2005; Lemonge and Barbosa, 2004; Bernardino et al., 2007; Coello and Cortes, 2004), the Branch and Bound (B&B) method (Ringertz, 1988), Particle Swarm Optimisation (PSO), Hybrid PSO (HPSO) and PSO with Passive Congregation (PSOPC) (Li et al., 2009), the Harmony Search (HS) algorithm (Lee et al., 2005), the Mine Blast Algorithm (MBA) (Sadollah et al., 2012), the Firefly Algorithm (FA) (Gandomi et al., 2011), Fuzzy Logic

(FL) (Shih and Yang, 2002), the nonlinear B&B Method (Sandgren, 1990), the extension of discrete steepest descent and rotating coordinate direction methods (Amir and Hasegawa, 1989), Evolutionary Algorithms (Eas) (Efren et al., 2003; Pant et al., 2009; Chen et al., 2009), the Bacterial Foraging Behaviour Algorithm (BFBA) (Montes et al., 2009), Adaptive Dimensional Search (ADS) (Hasançebi et al., 2015), Elitist Self-Adaptive Step-Size Search (ESASS) (Kazemzadeh et al., 2014) and the nonlinear B&B Method with Nonlinear Programming (NLP) Algorithm (Thanedar and Vanderplaats, 1995).

The performance of the CI-SPF method for solving structural and engineering domain problems is also compared with the MRSLS (Kulkarni et al., 2016) method. The MRSLS approach searches for an improved solution in the neighbourhood of the existing solution. The procedure can be described as follows:

> **Step 1:** A set of starting solutions X are randomly generated and the objective function $f(X) = f(X_1, X_2, ..., X_i, ..., X_n)$ is obtained.
>
> **Step 2:** All the variables $(X) = (X_1, X_2, ..., X_i, ..., X_n)$ are perturbed using a pairwise interchange approach with neighbourhood value $(X') = (X'_1, X'_2, ..., X'_1, ..., X'_n)$ and a neighbouring objective function $f(X')$ is generated.
>
> **Step 3:** If $f(X') \leq f(X)$ then accept $f(X')$ as the current solution X', i.e., if $f(X) = f(X')$; else retain $f(X)$. And return to Step 2.

This process continues until there is no significant change in the solution for a significant number of perturbations.

In the proposed work, the CI-SPF algorithm is coded in MATLAB 7.7.0 (R2011b) and the simulations are run on a Windows platform using an Intel(R) Core (TM) 2Duo, with a 2.93GHz processor speed and 4GB RAM. Furthermore, each problem is solved 30 times with different initial solutions. The mathematical formulation, results and comparison of the solutions with other contemporary algorithms are discussed in the following sections. The CI-SPF has been successfully applied in solving the discrete truss structure and mixed variable mechanical engineering design problems. Also, for a meaningful comparison, every MRSLS run is initialised to start in the neighbourhood of the starting point of one of the CI candidates in the CI-SPF algorithm.

3.2.1 Generalised Mathematical Formulation of Truss Structure Problems

The truss structure problems discussed in this book have discrete variables and aim to minimise weight (W) (see Eq. 3.3) subject to constraints on the maximum allowable stress (see Eq. 3.4) and displacement (see Eq. 3.5). For every element of the structure, the discrete variables are selected from within the predefined set of cross-sectional area A_i, ρ (weight density of material) and l_i (length of a truss member) are fixed. In general, the problem definition for the truss structure is stated as follows:

$$\text{Minimise } f = W = \sum_{i=1}^{N} \rho A_i l_i \tag{3.5}$$

$$\text{Subject to } |\sigma_i| \le \sigma_{max} \quad i = 1,2,...,P \tag{3.6}$$

$$|u_j| \le u_{max} \quad j = 1,2,...,M \tag{3.7}$$

where W   Objective function (Weight)

A_i Set of cross-sectional area of every truss structure member i, $i = 1,2,...,N$

ρ Weight density of the truss structure material

l_i Length of truss structure member i, $i = 1,2,...,N$

σ_{max} Maximum allowable stress

u_{max} Maximum allowable displacement

N Number of members in the truss structure

M Number of nodes in the truss structure

3.2.2 6-Bar Truss Structure Problem

The 6-bar truss structure (see Figure 3.3) problem was formerly discussed in Nanakorn and Meesomklin (2001). There are six design variables (cross-sectional area) equal to the number of members of the truss. The cross-sectional area of each member is taken from the set of discrete values $A_i \in$ {1.62, 1.80, 1.99, 2.13, 2.38, 2.62, 2.63, 2.88, 2.93, 3.09, 3.13, 3.38, 3.47, 3.55, 3.63, 3.84, 3.87, 3.88, 4.18, 4.22, 4.49, 4.59, 4.80, 4.97, 5.12, 5.74, 7.22, 7.97, 11.50, 13.50, 13.90, 14.20, 15.50, 16.00, 16.90, 18.80, 19.90, 22.00, 22.90, 26.50, 30.00, 33.50} in^2. The objective function is to minimise the weight satisfying tension and compression stress in each member and deflection at every node of the truss structure. The allowable stress is given as 25000 *psi*

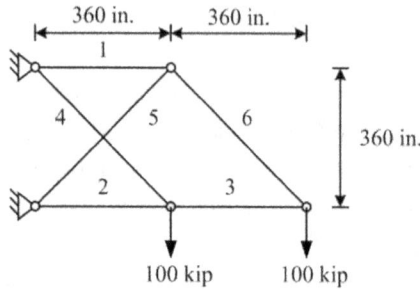

FIGURE 3.3   Planer 6-Bar Truss Structure.

TABLE 3.1   Comparison of Results for Solving the 6-Bar Truss Structure Problem.

Design Variables (in^2)	GA (Nanakorn and Meesomklin, 2001)	CI-SPF	MRSLS
A_1	30	30	30
A_2	19.9	19.9	19.9
A_3	15.5	15.5	15.5
A_4	7.22	7.22	7.22
A_5	22	22	22
A_6	22	22	22
Truss Weight $W(lb)$	4962.1	4962.0966	4962.0966
Function Evaluations	NA	3680	750

and allowable deflection is given as $2\,in$. The weight density of the material is $0.1\,lb/in^3$ and the modulus of elasticity is 10^7 psi. The problem definition for the truss structure is expressed in Eqs. 3.5, 3.6 and 3.7.

For 6-bar truss structure problem, the solutions obtained using CI-SPF and MRSLS are compared with GA (see Table 3.1). Statistical results such as the best, mean and worst function values obtained from 30 trials are same, i.e., 4962.0966 having standard deviation zero, the average number of function evaluations 3680 and average CPU time 4.58. Other statistical results, such as standard deviation of function evaluations, standard deviation CPU time, closeness to the reported solution and associated parameters are presented in Table 6.35. The convergence plot is presented in Figure 3.4. In the convergence plot, C1–C7 represents the number of cohort candidates used during the computation. From the comparison, it was noticed that CI-SPF obtained very similar results to GA.

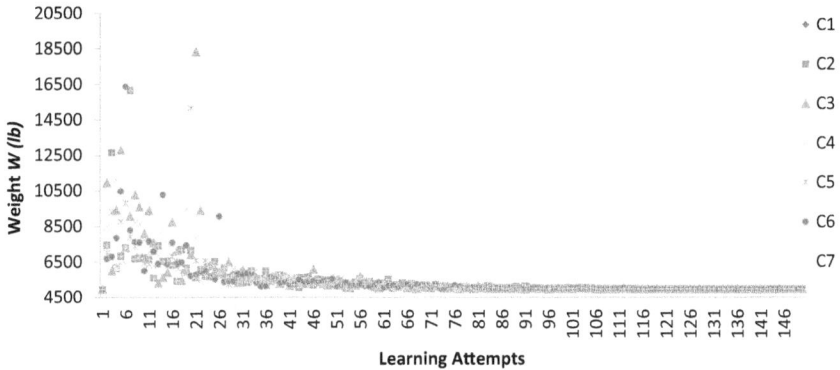

FIGURE 3.4   Convergence Plot for CI-SPF Solving the 6-Bar Truss Structure Problem.

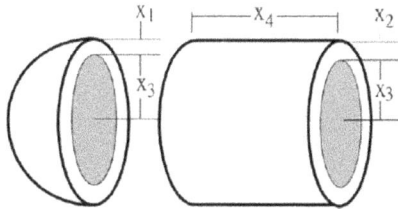

FIGURE 3.5   Tube and Pressure Vessel.

3.2.3 Pressure Vessel Design Engineering Problems

Consider the optimal design problem of a pressure vessel given in Sandgren (1990) depicted in Figure 3.5 where x_1 (the spherical head thickness) and x_2 (the shell thickness) are discrete variables, and x_3 (the radius of the shell) and x_4 (the length of the shell) are continuous variables. The detailed mathematical formulation is presented in the Appendix.

The best, mean and worst function values obtained using 30 trials of CI-SPF are 6059.7152, 6069.6323 and 6091.6584 with standard deviation 12.64, average CPU time 4.93 sec. The comparison of results is presented in Table 3.2. The convergence curve is presented in Figure 3.6. The other statistical details and parameters are presented in Table 6.35.

3.2.4 Linear and Nonlinear Test Problems

The integer variable linear programming problem and Rosen-Suzuki convex programming problem were previously solved by Srivastava and

TABLE 3.2 Comparison of Results for the Pressure Vessel Design Problem.

Variables	NIDPM (Sandgren, 1990)	Augmented Lagrange (Kannan and Kramer, 1994)	GA (Coello, 2000)	CPSO (He and Wang, 2007)	LCA (Kashan, 2011)	OIO (Kashan, 2015)	CI-SPF (Kale and Kulkarni, 2018)	MRSLS
x_1	1.125	1.125	0.8125	0.8125	NA	NA	0.8125	0.8125
x_2	0.625	0.625	0.4375	0.4375	NA	NA	0.4375	0.4375
x_3	48.97	58.291	40.3239	42.09126	NA	NA	42.0984	41.9645
x_4	106.72	43.69	200	176.7465	NA	NA	176.6366	178.3043
Cost $f(x)$	7981.5690	7198.0428	6288.7445	6061.0777	6059.8553	6059.7143	6059.7152	6076.1215
Function Evaluations	NA	NA	900000	200000	24000	50000	124581	1200

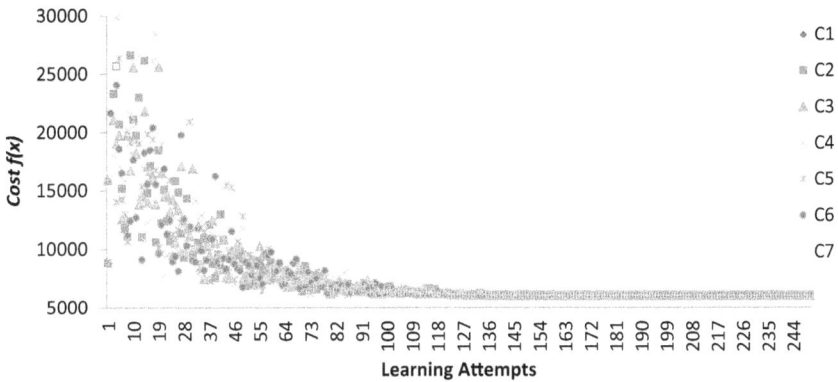

FIGURE 3.6   Convergence Curve for CI-SPF Solving the Pressure Vessel Design Problem.

Fahim (2001) using a modified gradient type optimisation technique. The linear programming problem was solved for maximisation; whereas, Rosen–Suzuki convex programming problem was solved for minimisation of function. The detailed mathematical formulation is presented the Appendix.

Apart from the truss structure and design engineering problems, linear and nonlinear problems are also solved using CI-SPF and MRSLS algorithms. The comparison of results is presented in Table 3.3. For the integer linear programming problem, MRSLS obtained better results as compared to CI-SPF and modified gradient type optimisation (Srivastava and Fahim, 2001), whereas Rosen–Suzuki, CI-SPF and MRSLS have obtained the same results as reported when using modified gradient type optimisation (Srivastava and Fahim, 2001). The convergence curves for integer linear programming problem and the Rosen–Suzuki problem are presented in Figures 3.7 and 3.8, respectively. The CI-SPF and MRSLS have obtained similar results for the 6-bar truss structure problem as compared to GA (Nanakorn and Meesomklin, 2001). For the pressure vessel problem, precisely similar results obtained by CI-SPF as compared to those of best reported solutions by LCA (Kashan, 2011) and Optical Inspired Optimisation (OIO) (Kashan, 2015) algorithms; however, the CI-SPF required a greater number of function evolutions as compared to LCA and OIO.

TABLE 3.3   CI-SPF Results from Solving the Linear and Nonlinear Problems.

| Sr. No. | Test Functions | Solver | Search Space | | Function Value | Optimum variables |
			Lower limit	Upper limit		
	Integer Linear Problem (Maximisation)	Srivastava and Fahim (2001)	[0,…,0]	[200, …,200]	316	[4,87,34,149,0]
		CI-SPF			1037	[200,199,67,104,0]
		MRSLS			1040	[200,200,67,106,0]
1	Rosen–Suzuki Test Problem Convex Programming Problem (Minimisation)	Srivastava and Fahim (2001)	[-10,-10,-10,-10]	[20, 20, 20, 20]	-44	[0,1,2,-1]
		CI-SPF			-44	[0,1,2,-1]
		MRSLS			-44	[0,1,2,-1]

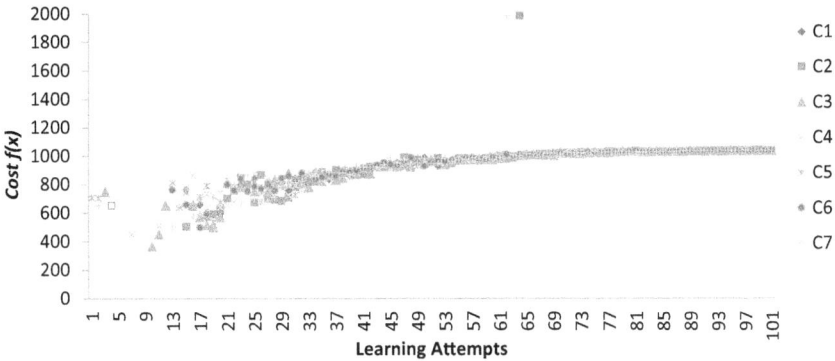

FIGURE 3.7 Convergence Curve for CI-SPF Solving the Integer Linear Programming Test Problem.

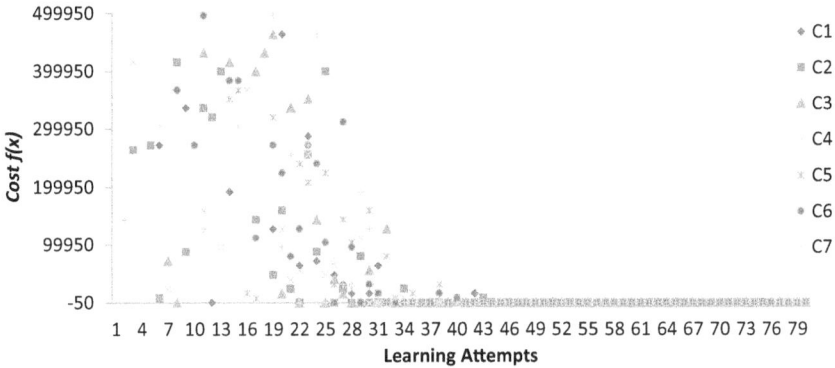

FIGURE 3.8 Convergence Curve for CI-SPF Solving the Rosen–Suzuki Test Problem.

3.3 ANALYSIS OF CI VARYING NUMBER OF CANDIDATES (C) AND SAMPLING SPACE REDUCTION FACTOR (R)

It has been observed from previous work (Krishnasamy et al., 2014; Kulkarni et al., 2013, 2015) as well as from the current work that in a similar way to other algorithms, the performance of CI is dependent on parameters such as the number of candidates C and the sampling space reduction factor R. The effect on the solution quality from varying the value of C is presented in Tables 3.4 and 3.6. Similarly, the effect on the solution by varying R is presented in Tables 3.5 and 3.7. For the analysis of C and R, 30 runs have been conducted for various values of $C = \{2,3,\ldots,10\}$ and $R = \{0.9,0.91,\ldots0.99\}$ solving representative discrete variable 10-bar truss structure problem and mixed variable stepped cantilever beam problem.

TABLE 3.4 Testing of the 10-Bar Truss Structure Problem Using CI-SPF for Various Values of C Keeping R Constant ($R = 0.97$).

Variables	C=2	C=3	C=4	C=5	C=6	C=7	C=8	C=9	C=10
A_1	33.5	33.5	33.5	33.5	33.5	33.5	33.5	33.5	33.5
A_2	1.62	1.62	1.62	1.62	1.62	1.62	1.62	1.62	1.62
A_3	22.9	22.9	22.9	22.9	22	22.9	22	22.9	22.9
A_4	14.2	14.2	16	13.9	16	14.2	16	13.9	14.2
A_5	1.62	1.62	1.62	1.62	1.62	1.62	1.62	1.62	1.62
A_6	1.62	1.62	1.62	1.62	1.8	1.62	1.62	1.62	1.62
A_7	7.97	7.97	7.97	7.97	7.22	7.97	7.22	7.97	7.97
A_8	22.9	22.9	22.9	22.9	22.9	22.9	22.9	22.9	22.9
A_9	22	22	19.9	22	22	22	22	22	22
A_{10}	1.62	1.62	3.13	1.62	1.62	1.62	1.62	1.62	1.62
Truss Weight $W(lb)$	5490.738	5490.738	5491.717	5491.614	5497.605	5490.738	5491.685	5491.614	5490.738
Time	3.82	6.21	6.38	7.32	11.70	7.97	14.74	13.16	19.98
FE	6306	6511	10584	12080	18090	14385	22112	21807	31330
Function STDV	48.99	27.58	46.32	24.38	37.05	37.82	33.05	40.86	24.74
Time STDV	1.45	3.28	3.58	3.97	5.49	4.42	6.07	9.48	8.96
FE STDV	2400.71	3442.85	5943.89	6463.79	8500.51	7973.81	9115.25	15316.22	14049.84

TABLE 3.5 Testing of the 10-Bar Truss Structure Problem Using CI-SPF for Various Values of R Keeping C Constant ($C = 7$).

Variables	R=0.90	R=0.91	R=0.92	R=0.93	R=0.94	R=0.95	R=0.96	R=0.97	R=0.98	R=0.99
A_1	33.5	33.5	30	33.5	33.5	33.5	33.5	30	33.5000	33.5000
A_2	2.38	1.8	1.62	2.13	1.99	2.13	1.8	1.62	1.6200	1.6200
A_3	22.9	22.9	22	22.9	22.9	22.9	22	22.9	22.9000	22.9000
A_4	16.9	15.5	14.2	13.5	15.5	13.9	16	16	15.5000	14.2000
A_5	1.62	1.62	1.62	1.62	1.62	1.62	1.62	1.62	1.6200	1.6200
A_6	3.13	1.8	2.88	1.8	1.62	1.62	1.99	1.62	1.6200	1.6200
A_7	7.97	11.5	11.5	7.97	7.97	7.97	7.97	7.97	7.9700	7.9700
A_8	22.9	19.9	22	22.9	22.9	22.9	22	22.9	22.0000	22.9000
A_9	19.9	19.9	22	22	19.9	22	22	22.9	22.0000	22.0000
A_{10}	2.62	3.87	2.62	2.88	1.99	1.62	1.62	1.62	1.6200	1.6200
Truss Weight $W(lb)$	5613.655	5610.473	5591.507	5564.979	5525.518	5507.557	5509.412	5497.709	5491.717	5490.737
Time	4.36	4.67	5.87	6.30	6.99	7.89	7.96	14.50	18.23	22.84
FE	7175	7924	9646	9464	11844	12915	12747	24115	24836	26754
Function STDV	44.53	46.86	40.75	46.25	58.13	29.74	38.51	47.49	25.87	23.95
Time STDV	1.93	2.35	1.82	2.66	2.80	3.40	4.56	4.84	5.97	7.37
FE STDV	3174.40	3999.30	2997.18	4004.61	4744.16	5570.09	7309.14	8051.13	8429.74	9747.56

TABLE 3.6 Testing of the Stepped Cantilever Beam Problem Using CI-SPF for Various Values of C Keeping R Constant ($R = 0.934$).

Variables	C=2	C=3	C=4	C=5	C=6	C=7	C=8	C=9	C=10
b_1	3	3	3	3	3	3	3	3	3
h_1	60	60	60	60	60	60	60	60	60
b_2	3.1	3.1	3.1	3.1	3.1	3.1	3.1	3.1	3.1
h_2	55	55	55	55	55	55	55	55	55
b_3	2.8	2.6	2.6	2.6	2.6	2.6	2.8	2.6	2.6
h_3	48	50	50	50	50	50	48	50	50
b_4	2.2046	2.2047	2.2046	2.2047	2.2047	2.2047	2.2048	2.206	2.2046
h_4	44.0915	44.0905	44.0912	44.0901	44.0896	44.0904	44.0916	44.0768	44.0909
b_5	1.7498	1.7498	1.7851	1.75	1.7498	1.7506	1.801	1.75	1.7501
h_5	34.996	34.9954	34.6487	34.9927	34.9959	34.9873	34.4945	34.9941	34.9922
Volume $f(cm^3)$	64334.129	63894.022	63955.472	63894.311	63894.215	63895.167	63894.236	63897.282	63894.473
Time	1.37	2.80	4.36	6.38	5.97	6.20	7.58	7.13	7.21
FE	5028	8532	12320	22350	21348	24920	23296	23319	22120
Function STDV	1241.64	1316.22	1361.16	625.34	533.91	240.48	228.83	240.77	139.06
Time STDV	0.88	1.24	2.258	1.76	2.59	2.10	3.94	4.14	3.46
FE STDV	3242.14	3796.29	6372.97	6186.54	9269.77	7988.59	12120.68	13531.06	10646.11

TABLE 3.7 Testing of the Cantilever Stepped Beam Problem Using CI-SPF for Various Values of R Keeping C Constant ($C = 7$).

Variables	R=0.90	R=0.91	R=0.92	R=0.93	R=0.94	R=0.95	R=0.96	R=0.97	R=0.98	R=0.99
b_1	3	3	3	3	4	3	3	3	3	3
h_1	60	60	60	60	55	60	60	60	60	60
b_2	3.1	3.1	3.1	3.1	3.1	3.1	3.1	3.1	3.1	3.1
h_2	55	55	55	55	55	55	55	55	55	55
b_3	2.6	2.6	2.6	2.8	2.6	2.6	2.6	2.6	2.6	2.6
h_3	50	50	50	48	50	50	50	50	50	50
b_4	2.2047	2.2046	2.2046	2.2048	3.0589	2.2047	2.2051	2.2046	2.2061	2.213
h_4	44.0897	44.0911	44.0911	44.0897	37.434	44.091	44.0889	44.0912	44.0846	44.1908
b_5	1.7499	1.752	1.7498	1.7501	1.7498	1.75	1.75	1.7503	1.7525	1.7705
h_5	34.9945	3	34.9951	34.9941	34.9959	34.9941	35.0003	34.9923	35.0365	34.8455
Volume $f(cm^3)$	63894.41	63897.44	63893.49	63898.28	63896.38	63894.47	63897.52	63895.15	63915.74	63998.83
Time	2.99	5.33	4.84	6.82	6.86	7.58	11.23	11.90	13.10	40.97
FE	12523	18627	20174	20783	22764	25165	37247	39480	43470	75922
Function STDV	704.24	925.84	606.06	237.94	253.60	257.78	235.97	236.15	221.56	203.95
Time STDV	1.53	1.19	2.39	3.70	3.09	3.57	5.52	6.11	6.49	15.68
FE STDV	6409.10	4180.69	9951.64	11275.74	10259.07	11849.03	18320.68	20272.44	21545.25	29060.99

For the discrete variable 10-bar truss structure problem and the mixed variable stepped cantilever beam problem the variation in the function value, computational time and function evaluations along with their standard deviations are listed in Table 3.4 and Table 3.6, respectively. It is noted that with a fewer number of candidates, i.e., $C = 2$ or $C = 3$ keeping the values of $R = 0.97$ (10-bar truss structure problem) and $R = 0.932$ (stepped cantilever beam problem) fixed, the required average computational time, function evaluations, and standard deviations are considerably lower with better objective function value. However, the solution may not be acceptable because the standard deviation of the function value is too large. This may be because the considerable number of choices available to follow the best are fewer. On the other hand, for a greater number of candidates the number of choices also increases which helps to improve cohort behaviour. Additionally, it is observed that the time and function evaluations also increased linearly. However, as demonstrated in Figure 3.9 (c) and (f), the function values as well as the associated standard deviations do not improve significantly, whereas in mixed variable stepped cantilever beam problem (see Figure 3.10) the trend shows that the standard deviation continues decreasing as the number of choices to follow the best candidate increases. From Table 3.4, it was noted that the best function value was obtained for $C = 7$.

The analysis of the values of the most effective parameter R is carried out with a constant value of $C = 7$ discussed in Table 3.5 for 10-bar truss structure problem and Table 3.7 for stepped cantilever beam problem. Moreover, the premature convergence of the solution is observed due to the smaller value of R (i.e., $R < 0.9$), which is not able to satisfy the available constraints. Thus, to avoid the premature convergence, a value of $R = \{0.9, 0.91,...0.99\}$ was selected on the basis of preliminary trials. As the value of the sampling space reduction factor R is increased, the solution quality improved; however, it eventually increased the computational time (Figure 3.11 (a)) and function evaluations (Figure 3.11 (b)). In contrast, in the case of the mixed variable stepped cantilever beam problem, if the value of R is gradually increased the reported time and function evaluations also increased; however, the function value worsened with improved standard deviation (Figure 3.12 (f)).

In this chapter, the investigation of CI-SPF algorithm for solving the 6-bar truss structure problem, the pressure vessel design engineering problem, the integer linear programming problem and the nonlinear problem

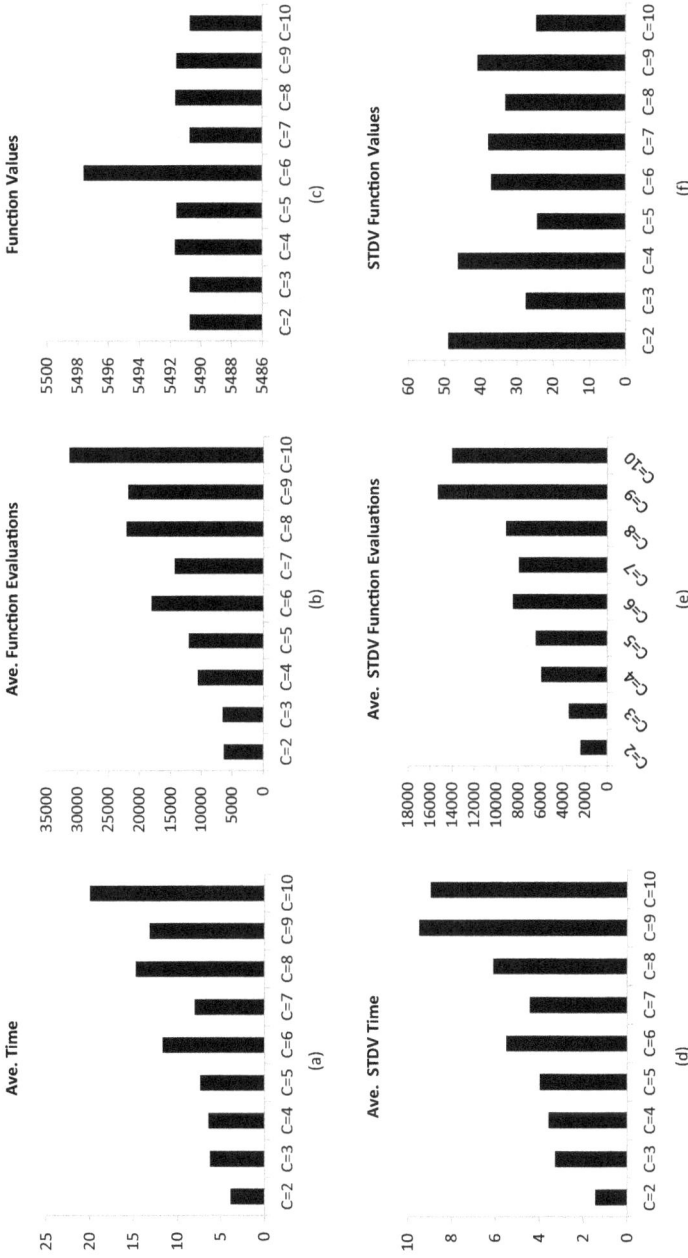

FIGURE 3.9 Testing of the 10-Bar Truss Structure Problem Using CI-SPF for Various Values of *C* Keeping *R* Constant (*R* = 0.97).

Note: (a) Average CPU Time, (b) Average number of Function Evaluations, (c) Function Value, (d) Average Standard Deviation of Time, (e) Average Standard Deviation of Function Evaluations and (f) Standard Deviation of Function Value.

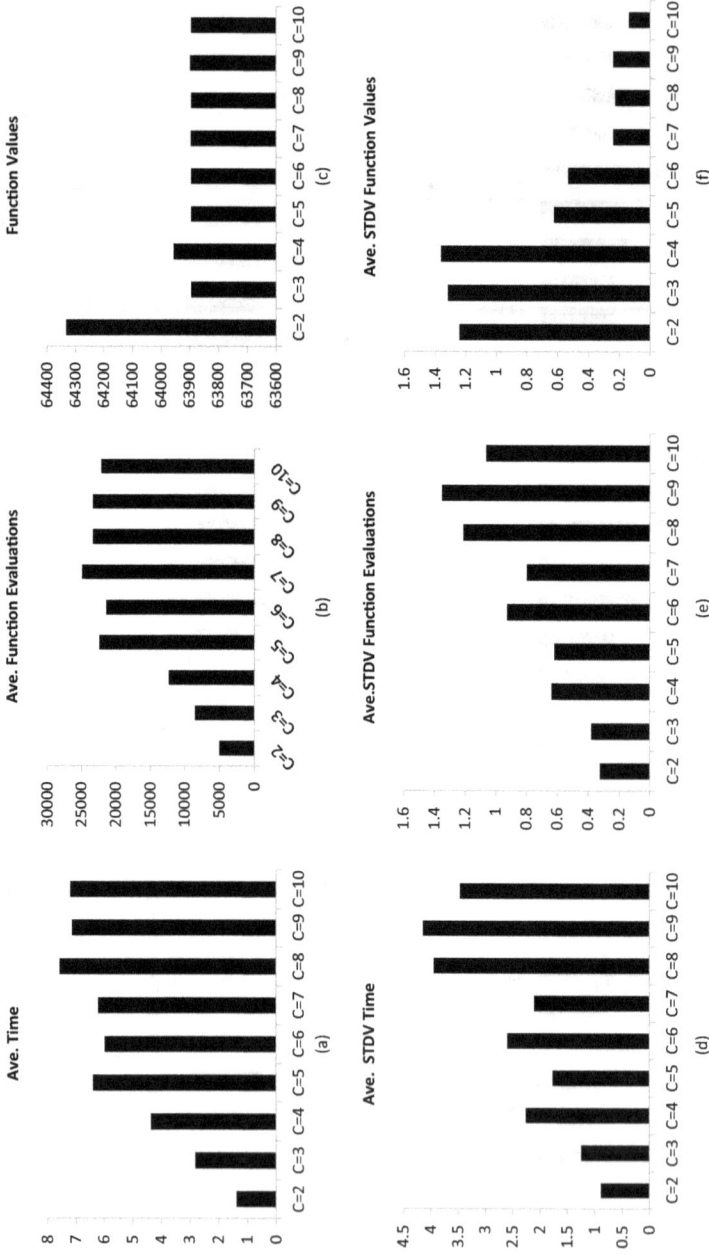

FIGURE 3.10 Testing of the 10-Bar Truss Structure Problem Using CI-SPF for Various Values of *R* Keeping *C* Constant (*C*=7).

Note: (a) Average CPU Time, (b) Average number of Function Evaluations, (c) Function Value, (d) Average Standard Deviation of Time, (e) Average Standard Deviation of Function Evaluations and (f) Standard Deviation of Function Value.

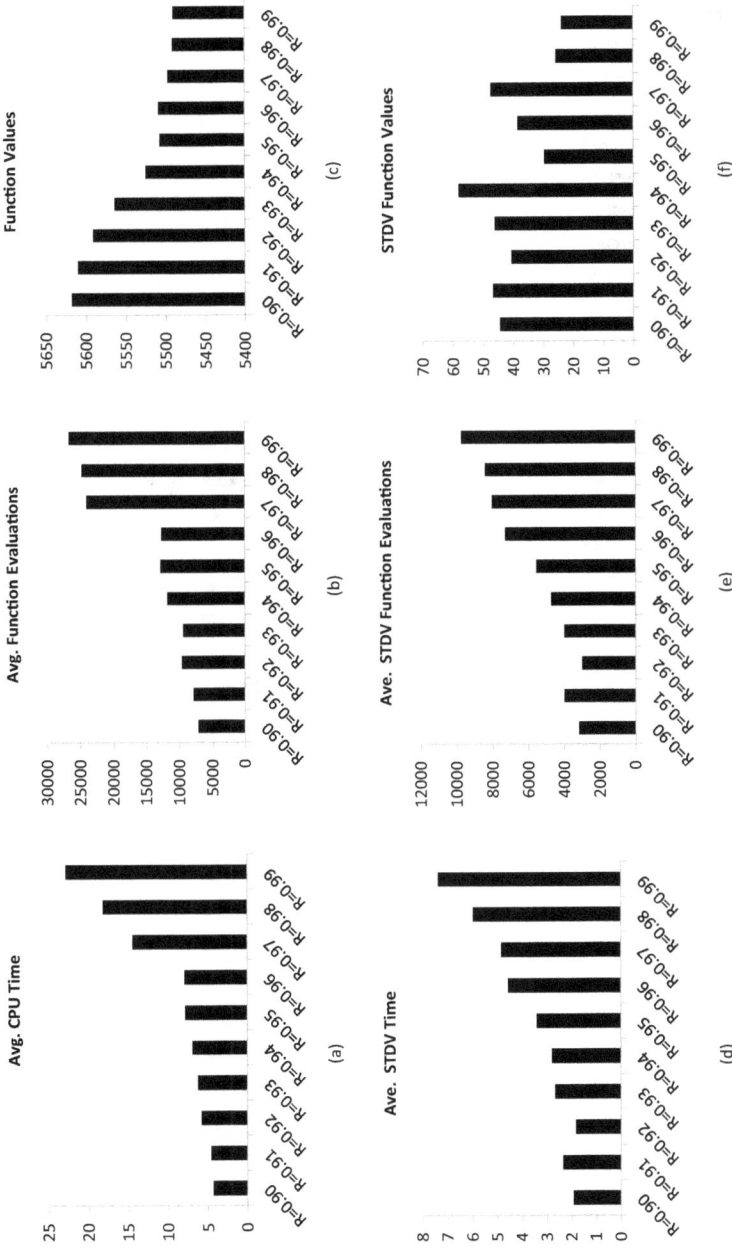

FIGURE 3.11 Testing of the Stepped Cantilever Beam Problem Using CI-SPF for Various Values of C Keeping *R* Constant (*R* = 0.934).

Note: (a) Average CPU Time, (b) Average number of Function Evaluations, (c) Function Value, (d) Average Standard Deviation of Time, (e) Average Standard Deviation of Function Evaluations and (f) Standard Deviation of Function Value.

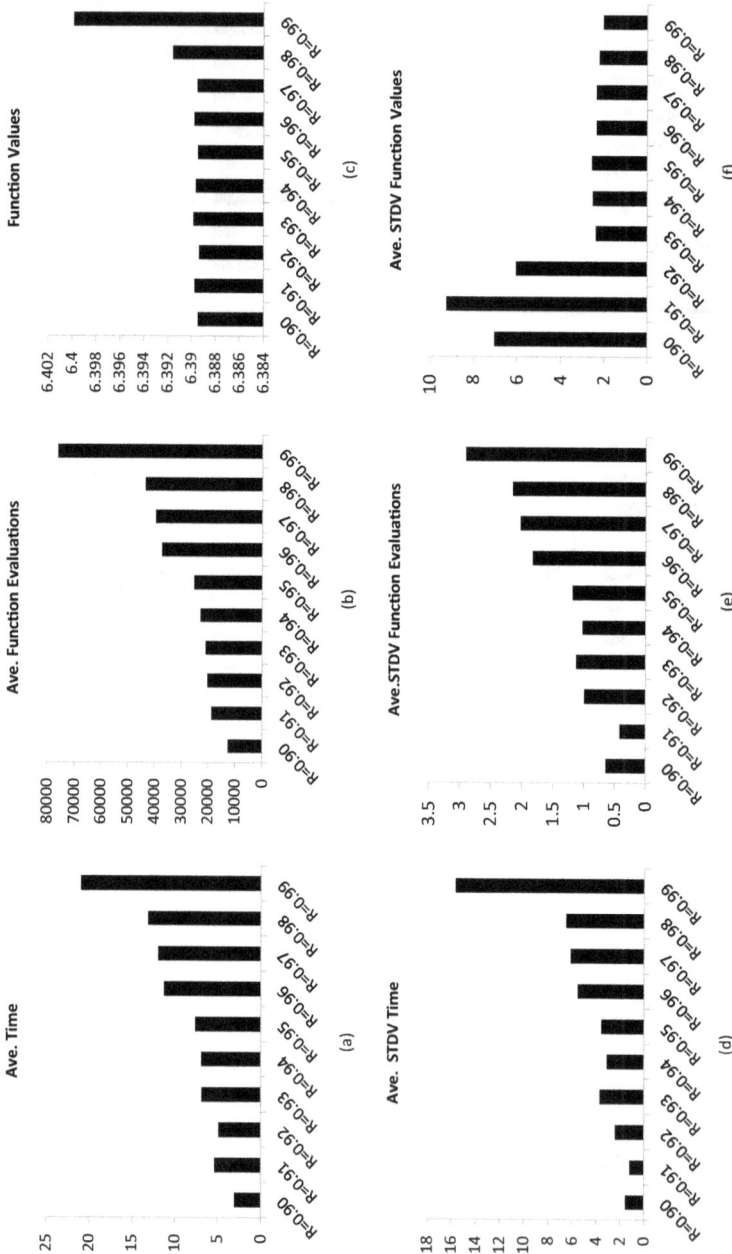

FIGURE 3.12 Testing of the Stepped Cantilever Beam Problem Using CI-SPF for Various Values of R Keeping C Constant ($C = 7$).

Note: (a) Average CPU Time, (b) Average number of Function Evaluations, (c) Function Value, (d) Average Standard Deviation of Time, (e) Average Standard Deviation of Function Evaluations and (f) Standard Deviation of Function Value.

are discussed. The remaining problems are illustrated and discussed in Chapter 6. For constraint handling, a well-known SPF approach was adopted as it is simple to implement. However, a limitation was observed that the SPF approach required a significant number of trials to set the associated penalty parameter. So in order to overcome this limitation, a SAPF approach is proposed (see Chapter 4). For the SAPF approach, the required penalty parameter is obtained using a CI algorithm. It does not require any parameter tuning. The approach is validated by solving above discussed problems.

3.4 CONCLUSION

For the first time, a CI algorithm was successfully applied in order to solve the 10-bar discrete variable truss structure problem and mixed variable mechanical engineering pressure vessel design problems. Constrained test problems such as the linear programming problem and the Rosen–Suzuki convex programming problem were also solved using CI. The discrete/integer variables included in these problems were handled using a round-off integer sampling technique. The CI algorithm was supplemented by using the penalty function approach as a constrained handling technique. The obtained results were analysed, and indicated that the CI approach was sufficiently robust with reasonable computational time period and a smaller number of function evaluations as compared to other contemporary methods. In addition, a MRSLS method was applied for solving structural and mechanical engineering problems; however, the results obtained using CI were more prominent than MRSLS method. Apart from these intrinsic worth properties, some downside properties were also identified. It was observed that quality performance of CI was dependent on the most effective parameters such as the number of candidates C, sampling space reduction factor R, number of variations t and penalty parameter θ. The parameters were derived empirically over numerous experiments and their calibration required some preliminary trials. Additionally, it was observed that as the problem size increased the computational time and function evaluations increased considerably. In the near future, it is intended that modifications be made to CI by using the uniform distributed probability (UDP) and applying for solving the linear and nonlinear discrete problems and large sized truss structural and mechanical engineering assembly design problems.

REFERENCES

Amir, H.M. and Hasegawa, T. (1989) 'Nonlinear mixed-discrete structural optimization', *Journal of Structural Engineering*, Vol. 115, No. 3, pp. 626–645.

Bernardino, H.S., Barbosa, H.J.C. and Lemonge, A.C.C. (2007) 'A hybrid genetic algorithm for constrained optimization problems in mechanical engineering', *Proceedings of IEEE Congress Evolution Computations*, pp. 646–653.

Chen, T.Y. and Chen, H.C. (2009) 'Mixed-discrete structural optimization using a rank-niche evolution strategy', *Engineering Optimization*, Vol. 41, No. 1, pp. 39–58.

Coello, C.A.C. (2000) 'Use of a self-adaptive penalty approach for engineering optimization problems', *Computer in Industry*, Vol. 41, pp. 113–127.

Coello, C.A.C. and Cortes, N.C. (2004) 'Hybridizing a genetic algorithm with an artificial immune system for global optimization', *Engineering Optimization*, Vol. 36, No. 5, pp. 607–634.

Deb, K. and Goyal, M. (1996) 'A combined genetic adaptive search (GeneAS) for engineering design', *Computer Science and Informatics*, Vol. 26, pp. 30–45.

Efren, M., Coello, C.A.C. and Ricardo, L. (2003) 'Engineering optimization using a simple evolutionary algorithm', *Proceedings of 15th International Conference on Tools with Artificial Intelligence (ICTAI)*, pp. 149–156.

Gandomi, A.H., Yang, X-S. and Alavi, A.H. (2011) 'Mixed variable structural optimization using firefly algorithm', *Computers and Structures*, Vol. 89, Nos. 23–24, pp. 2325–2336.

Goldberg, D.E. (1989) 'Genetic Algorithms', in *Search, Optimization and Machine Learning*, Addison-Wesley.

Hasançebi, O. and Kazemzadeh Azad, S. (2015) 'Adaptive Dimensional Search: A new metaheuristic algorithm for discrete truss sizing optimization', *Computers and Structures*, Vol. 154, pp. 1–16.

He, Q. and Wang, L. (2007) 'An effective co-evolutionary particle swarm optimization for constrained engineering design problem', *Engineering Applications of Artificial Intelligence*, Vol. 20, No. 1, pp. 89–99.

Kale, I.R. and Kulkarni, A.J. (2018) 'Cohort intelligence algorithm for discrete and mixed variable engineering problems', *International Journal of Parallel, Emergent and Distributed Systems*, Vol. 33, No. 6, pp. 627–662.

Kannan, B.K. and Kramer, S.N. (1994) 'An augmented Lagrange multiplier based method for mixed integer discrete continuous optimization and its applications to mechanical design', *Journal of Mechanical Design*, Vol. 116, No. 2, pp. 405–411.

Kashan, A.H. (2011) 'An efficient algorithm for constrained global optimization and application to mechanical engineering design: League championship algorithm (LCA)', *Computer-Aided Design*, Vol. 43, pp. 1769–1792.

Kashan A.H. (2015) 'A new metaheuristic for optimization: Optics inspired optimization (OIO)', *Computers and Operations Research*, Vol. 55, pp. 99–125.

Kazemzadeh Azad, S. and Hasançebi, O. (2014) 'An elitist self-adaptive step-size search for structural design optimization', *Applied Soft Computing*, Vol. 19, pp. 226–235.

Krishnasamy, G., Kulkarni A.J. and Paramesran R. (2014) 'A hybrid approach for data clustering based on modified cohort intelligence and K-means', *Expert Systems with Applications*, Vol. 41, pp. 6009–6016.

Kulkarni, A.J., Durugkar I.P. and Kumar M. (2013) 'Cohort intelligence: A self supervised learning behavior', *Systems, Man, and Cybernetics (SMC), IEEE International Conference*, pp. 1396–1400.

Kulkarni, A.J., Tai, K. and Abraham, A. (2015) Probability Collectives: A Distributed Multi-Agent System Approach for Optimization, Springer.

Kulkarni, A.J., Baki, M.F. and Chaouch, B.A. (2016) 'Application of the cohort-intelligence optimization method to three selected combinatorial optimization problems', *European Journal of Operational Research*, Vol. 250, No. 2, pp. 427–447.

Lee, K.S., Geem, Z.W., Lee, S.H. and Bae, K.W. (2005) 'The harmony search heuristic algorithm for discrete structural optimization', *Engineering Optimization*, Vol. 37, No. 7, pp. 663–684.

Lemonge, A.C.C. and Barbosa, H.J.C. (2004) 'An adaptive penalty scheme for genetic algorithms in structural optimization', *International Journal of Numerical Methods and Engineering*, Vol. 59, No. 5, pp. 703–736.

Li, L.J., Huang, Z.B. and Liu, F. (2009) 'A heuristic particle swarm optimization method for truss structures with discrete variables', *Computers and Structures*, Vol. 87, Nos. 7–8, pp. 435–443.

Montes, E.M. and Ocana, B.H. (2009) 'Modified bacterial foraging optimization for engineering design', in C.H. Dagli et al. (Eds) *Artificial Neural Networks in Engineering Conference (ANNIE), Intelligent Engineering Systems through Artificial Neural Networks*, Vol. 19, pp. 357–364.

Nanakorn, P. and Meesomklin, K. (2001) 'An adaptive penalty function in genetic algorithms for structural design optimization', *Computer and Structures*, Vol. 79, pp. 2527–2539.

Pant, M., Thangaraj, R. and Singh, V.P. (2009) 'Optimization of mechanical design problems using improved differential evolution algorithm', *International Journal of Recent Trends in Engineering*, Vol. 1, No. 5, pp. 21–25.

Rajeev, S. and Krishnamoorthy, C.S. (1992) 'Discrete optimization of structures using genetic algorithm', *Journal of Structural Engineering*, Vol. 118, No. 5, pp. 1123–1250.

Ringertz, U.T. (1988) 'On methods for discrete structural constraints', *Engineering Optimization*, Vol. 13, No. 1, pp. 47–64.

Sadollah, A., Bahreininejad, A., Eskandar, H. and Hamdi, M. (2012) 'Mine blast algorithm for optimization of truss structures with discrete variables', *Computers and Structures*, Vol. 49, No. 63, pp. 102–103.

Sandgren, E. (1990) 'Nonlinear integer and discrete programming in mechanical design optimization', *Journal of Mechanical Design*, Vol. 112, No. 2, pp. 223–229.

Shih, C.J. and Yang, Y.C. (2002) 'Generalized Hopfield network based structural optimization using sequential unconstrained minimization technique with additional penalty strategy', *Advances in Engineering Software*, Vol. 33, Nos. 7–10, pp. 721–729.

Srivastava, V.K. and Fahim, A. (2001) 'A two-phase optimization procedure for integer programming problems', *Computers and Mathematics with Applications*, Vol. 42, pp. 1585–1595.

Thanedar, P.B. and Vanderplaats, G.N. (1995) 'Survey of discrete variable optimization for structural design', *Journal of Structural Engineering*, Vol. 121, No. 2, pp. 301–306.

Wu, S.J. and Chow, P.T. (1995) 'Steady-state genetic algorithms for discrete optimization of trusses', *Computers and Structures*, Vol. 56, No. 6, pp. 979–991.

Yun, Y.S. (2005) 'Study on Adaptive Hybrid Genetic Algorithm and Its Applications to Engineering Design Problems', MSc thesis, Waseda University.

CHAPTER 4

Constraint Handling Using the Self-Adaptive Penalty Function (SAPF) Approach

4.1 INTRODUCTION

The mechanical design engineering and truss structure optimisation domain problems generally consist of discrete and mixed variables. These variables have very limited search space. Such problems become more difficult to solve when constraints are involved. In order to deal with the constraints, several constraint handling techniques have been proposed such as penalty function methods, probability distribution, feasibility-based rule. A detailed literature review on constraint handling is discussed in Chapter 2 (see Section 2.2). Initially, the SPF approach was incorporated in the CI algorithm in order to solve the constraint problems (see Chapter 3); however, SPF is dependent on the penalty parameter, which requires several trials to set a suitable value. This may increase the initial computational efforts.

In this chapter, in order to make the penalty function approach more effective the Self-Adaptive Penalty Function (SAPF) approach is proposed and incorporated in the CI algorithm. This approach eliminates the effort required in setting the penalty parameter and no other supporting parameter is required. This process means the CI-SAPF algorithm does not require the setting of penalty parameters, which may further help to generalise the algorithm.

4.2 SELF-ADAPTIVE PENALTY FUNCTION (SAPF)

In general, the constrained optimisation problem is expressed as follows:

$$\text{Minimize } f(X) = f(X_1, X_2, \dots, X_i, \dots, X_n) \tag{4.1}$$

subject to

$$g_i(X) \leq 0, \quad i = 1, 2, \dots, p$$

$$h_i(X) = 0, \quad i = 1, 2, \dots, m$$

$$\Psi^{lower} \leq X \leq \Psi^{upper}$$

As constraints are available with the problem a static penalty function (SPF) constraint handling approach was widely used. It is expressed as follows:

$$PF = \theta \times \left(\sum_{i=1}^{p} g_i(X) + \sum_{i=1}^{m} h_i(X) \right) \tag{4.2}$$

where θ is a penalty parameter and $\left(\sum_{i=1}^{p} g_i(X) + \sum_{i=1}^{m} h_i(X) \right)$ is the summation of the violated constraints. However, the θ value needs to be chosen arbitrarily to penalise the violated solution. Use of SPF necessitates several preliminary trials to set a suitable θ value. This is the limitation SPF approach. In order to overcome this limitation in setting the value of θ, an SAPF is proposed. In SAPF, the objective function $f(X)$ is utilised as a penalty parameter expressed as follows:

$$SAPF = f(X) \times \left(\sum_{i=1}^{p} g_i(X) + \sum_{i=1}^{m} h_i(X) \right) \tag{4.3}$$

And further forms the pseudo objective function $\phi(X)$ as follows:

$$\phi(X) = f(X) + SAPF \tag{4.4}$$

Equation (4.4) is only applicable for a positive function value, i.e., $f(X)$. There are other applications (numerical problems) with negative function values, i.e., $-f(X)$; for those, SAPF is expressed as follows:

$$SAPF = abs\left(f\left(X\right)\right)\times\left(\sum_{i=1}^{p}g_{i}\left(X\right)+\sum_{i=1}^{m}h_{i}\left(X\right)\right) \qquad (4.5)$$

Furthermore, the pseudo objective function $\phi(X)$ is formed as follows:

$$\phi\left(X\right)=-f\left(X\right)+SAPF \qquad (4.6)$$

Specifically, when the function value of any of the problems is too small or zero, the SAPF approach may not give a feasible solution. In such cases, an arbitrary integer (fixed) value (*int*) is added in the function value $f(X)$. Then the SAPF is calculated as follows:

$$SAPF = \left(f\left(X\right)+int\right)\times\left(\sum_{i=1}^{n}g_{i}\left(X\right)+\sum_{i=1}^{m}h_{i}\left(X\right)\right) \qquad (4.7)$$

The experimental analysis of CI-SAPF algorithm is discussed in Section 4.2.

4.3 CI-SAPF

Consider a cohort with number of candidates C. For every individual candidate $c\,(c=1,2,\ldots,C)$, using the CI-SAPF approach the pseudo objective function (behaviour) can be written as follows:

$$\phi\left(X^{c}\right)=f\left(X^{c}\right)+SAPF\left(X^{c}\right) \qquad (4.8)$$

where $SAPF\left(X^{c}\right)=f\left(X^{c}\right)\times\left(\sum_{i=1}^{p}g_{i}\left(X^{c}\right)+\sum_{i=1}^{m}h_{i}\left(X^{c}\right)\right)$ is penalty function and $f(X^{c})$ is the objective function of an individual candidate. The behaviour of the penalty parameter is dependent on the variable sampling space because, as the algorithm progresses it narrows down the sampling space using a sampling space reduction factor. An independent penalty parameter $f(X^{c})$ is generated by every individual candidate c to penalise the cohort behaviour (function value) with violated constraints. Additionally, the penalty parameter is updated for every learning attempt of CI-SAPF algorithm. The CI-SAPF algorithm was run for 30 times for each problem. In this chapter, the experimentation of CI-SAPF algorithm was carried out for solving one problem from each domain, i.e., the discrete variable 6-bar truss structure problem, the mixed variable pressure vessel design engineering problem, the discrete variable linear programming problem and the Rosen–Suzuki nonlinear convex test problem. The solutions to rest of the problems are discussed in Chapter 6.

4.4 PROBLEMS SOLVED

4.4.1 6-Bar Truss Structure Problem

For the 6-bar truss structure problem, the CI-SAPF algorithm achieved the same results as compared to GA (see Table 4.1). The best, mean and worst solutions obtained from 30 trials using the CI-SAPF algorithm are the same ($4962.0966\,lb$) with zero standard deviation and average computational time $4.58\,sec$. The average function value reported using GA was $5250\,lb$. From this, it was observed that the CI-SAPF performed marginally better than the GA. The average number of function evaluations was 2735. The convergence plot for CI-SAPF is presented in Figure 4.1(a). The other computational details of CI-SAPF associated with this problem are presented in Table 6.35.

4.4.2 Pressure Vessel Design Engineering Problem

The pressure vessel design problem for cost minimisation was successfully solved using CI-SAPF. The results obtained using the CI-SAPF algorithm compared with other contemporary algorithms are presented in Table 4.2. The best, mean and worst CI-SAPF solutions obtained from 30 trials were 6051.4819, 6065.2506 and 6103.2707 with standard deviation 11.9107,

TABLE 4.1 Comparison of Results for Solving the 6-Bar Truss Structure.

Design Variables (in^2)	GA (Nanakorn and Meesomklin, 2001)	CI-SPF	MRSLS	CI-SAPF
A_1	30	30	30	30
A_2	19.9	19.9	19.9	19.9
A_3	15.5	15.5	15.5	15.5
A_4	7.22	7.22	7.22	7.22
A_5	22	22	22	22
A_6	22	22	22	22
Truss Weight W (lb)	4962.1	4962.0966	4962.0966	4962.0966
Function Evaluations	NA	3680	750	2250

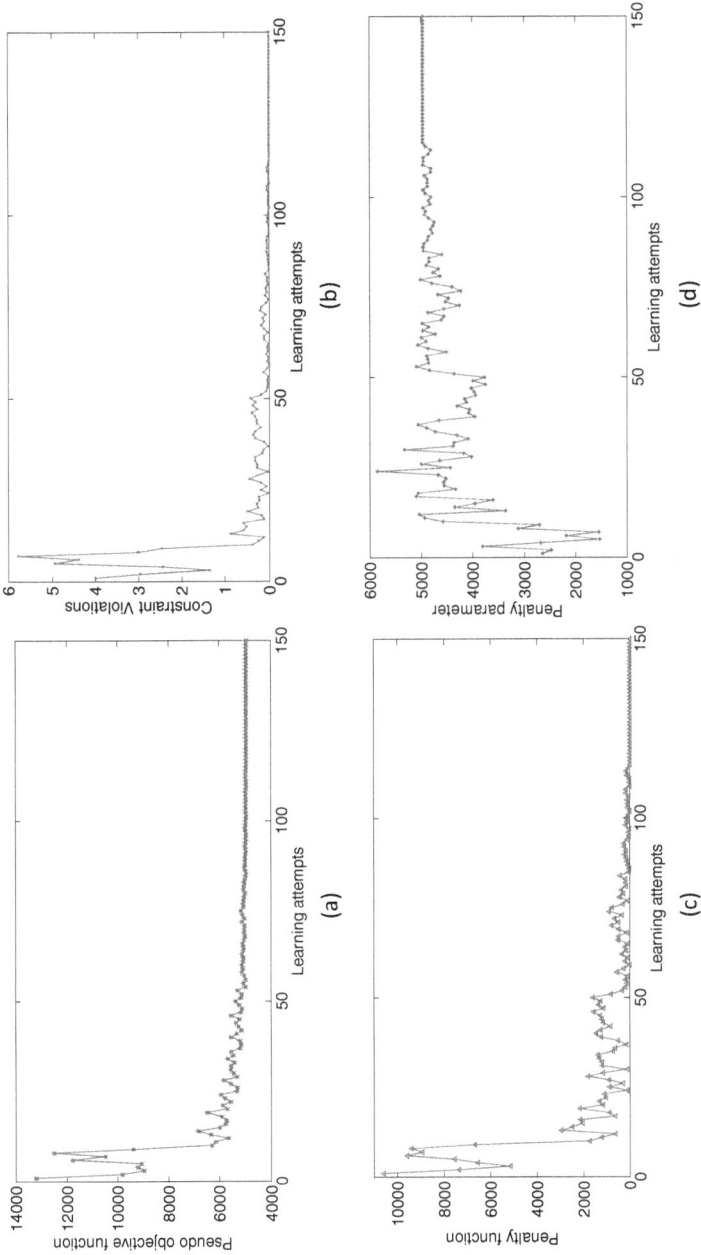

FIGURE 4.1   Behaviour of the Pseudo Objective Function (a), Constraint Violations (b), Penalty Function (c) and Penalty Parameter (d) Solving Discrete Variable 6-Bar Truss Structure Problem Using CI-SAPF.

average CPU time 4.43 *sec* and average number of function evaluations 11974. The convergence curve for CI-SAPF is presented in Figure 4.2 (a).

4.4.3 Linear and Nonlinear Test Problems

The benchmark linear and nonlinear test problems are adopted from Srivastava and Fahim (2001). These problems consist of integer variables and nonlinear constraints. These problems were solved using a two-phase optimisation procedure. A comparison of solutions obtained from the CI-SAPF algorithm and other contemporary techniques is presented in Table 4.3. When solving the nonlinear Rosen–Suzuki convex programming problem, the CI-SAPF obtained similar results. The CI-SAPF obtained better results when solving integer linear programming problems than the two-phase optimisation procedure and the gradient-based steepest descent and hem-stitching approaches (Srivastava and Fahim, 2001).

In CI-SAPF, the penalty parameter (objective function) is generated iteratively in a way that depends on the available set of design variables (Figures 4.1–4.4 (d)). The impact of this penalty parameter on the behaviour of penalty function (Figures 4.1–4.4 (c)), constraint violations (Figures 4.1–4.4 (b)) and pseudo objective function (Figures 4.1–4.4 (a)) is observed to be similar to the calculated penalty parameter. The CI-SAPF is separately tested on four problems from different domains. The graphical representation of the discrete variable 6-bar truss structure (Figure 4.1), mixed variable pressure vessel design engineering (Figure 4.2), discrete variable linear (Figure 4.1) and Rosen-Suzuki nonlinear convex (Figure 4.4) problems are presented. As the penalty parameter value is iteratively updated, it behaves like a dynamic penalty function, and a similar trend is obtained from the pseudo objective function, which represents its ability to come out from the local minima and help to avoid the premature convergence of the solution.

4.5 CONCLUSION

The CI-SAPF has been successfully validated for solving discrete/integer and mixed variable problems, as discussed in Section 4.4. The proposed SAPF approach is easy to apply and does not require the parameter tuning of the SPF approach. The CI-SAPF approach obtained similar results to CI-SPF within a smaller number of function evaluations. The effect of the SAPF approach on the pseudo objective function, constraint violations and penalty parameter are also discussed in detail. During the computation

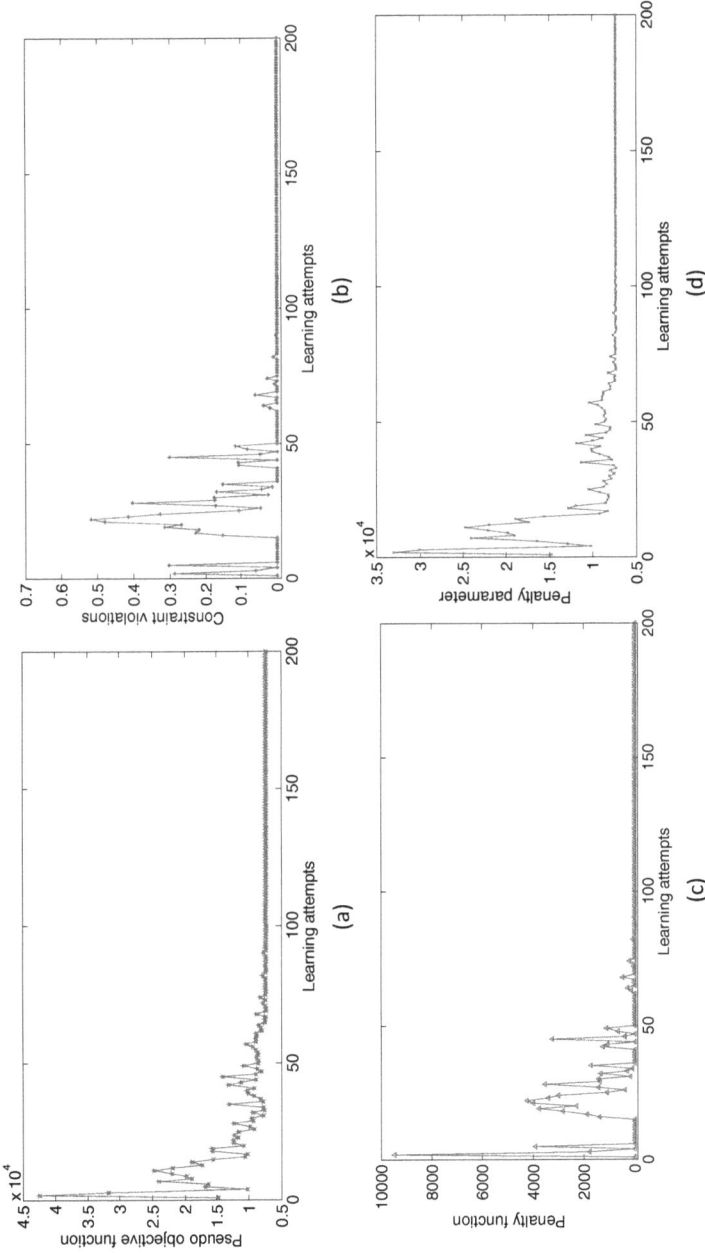

FIGURE 4.2   Behaviour of Pseudo Objective Function (a), Constraint Violations (b), Penalty Function (c) and Penalty Parameter (d) Solving Mixed Variable Pressure Vessel Problem Using CI-SAPF.

TABLE 4.2 Comparison of Results for the Pressure Vessel Design Problem.

Variables	NIDPM (Sandgren, 1990)	Augmented Lagrange (Kannan and Kramer, 1994)	GA (Coello, 2000)	CPSO (He and Wang, 2007)	LCA (Kashan, 2011)	OIO (Kashan, 2015)	MRSLS	CI-SPF (Kale and Kulkarni, 2018)	CI-SAPF
x_1	1.125	1.125	0.812500	0.812500	NA	NA	0.8125	0.8125	0.75
x_2	0.625	0.625	0.437500	0.437500	NA	NA	0.4375	0.4375	0.375
x_3	48.97	58.29100	40.32390	42.09126	NA	NA	41.9645	42.0984	40.437
x_4	106.72	43.6900	200.0000	176.7465	NA	NA	178.3043	176.6366	198.3782
Cost $f(x)$	7981.5690	7198.0428	6288.7445	6061.0777	6059.8553	6059.7143	6076.1215	6059.7152	6051.4819
Function Evaluations	NA	NA	900000	200000	24000	50000	1200	124581	9744

TABLE 4.3 CI-SAPF Results When Solving Linear and Nonlinear Problems.

Sr. No.	Test Functions	Solver	Search Space Lower limit	Search Space Upper limit	Function Value	Optimum variables
1	Integer Linear Problem (Maximisation)	Srivastava and Fahim (2001)	[0, ...,0]	[200, ...,200]	316	[4,87,34,149,0]
		CI-SPF			1037	[200,199,67,104,0]
		MRSLS			1040	[200,200,67,106,0]
		CI-SAPF			1037	[200,199,67,104,0]
2	Rosen-Suzuki Test Problem Convex Programming Problem (Minimisation)	Srivastava and Fahim (2001)	[-10,-10,-10,-10]	[20, 20, 20, 20]	-44	[0,1,2,-1]
		CI-SPF			-44	[0,1,2,-1]
		MRSLS			-44	[0,1,2,-1]
		CI-SAPF			-44	[0,1,2,-1]

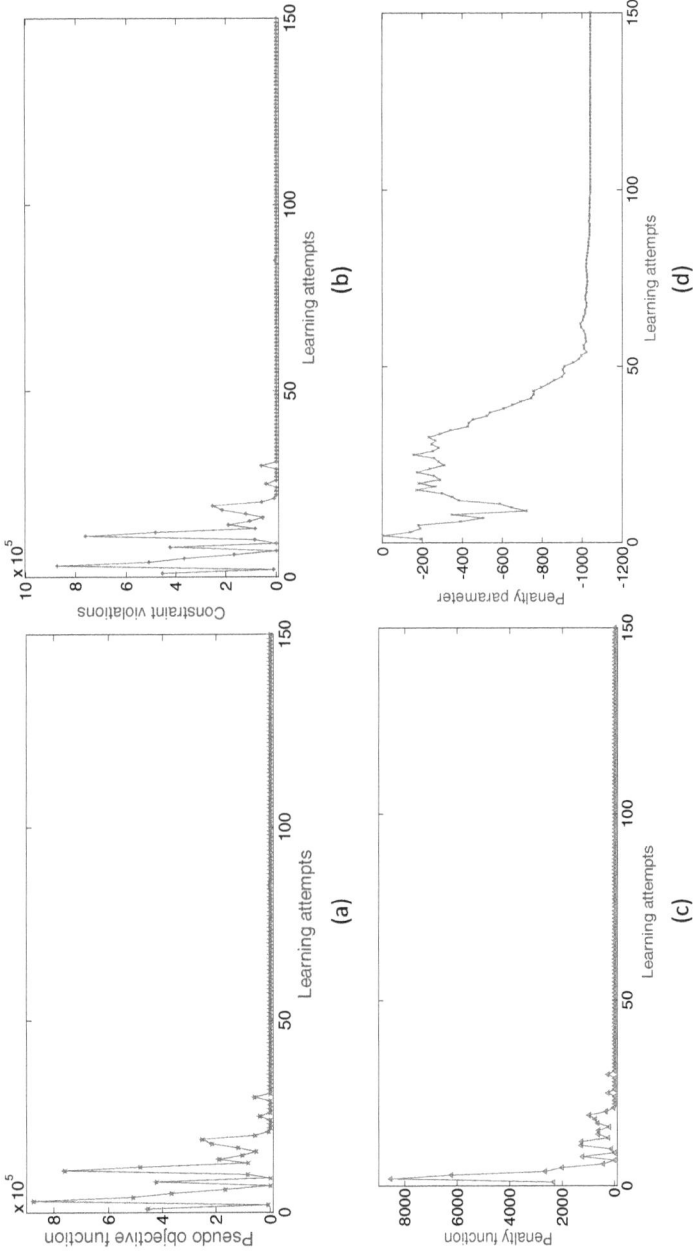

FIGURE 4.3 Behaviour of Pseudo Objective Function (a), Constraint Violations (b), Penalty Function (c) and Penalty Parameter (d) Solving Linear Programming Problem Using CI-SAPF.

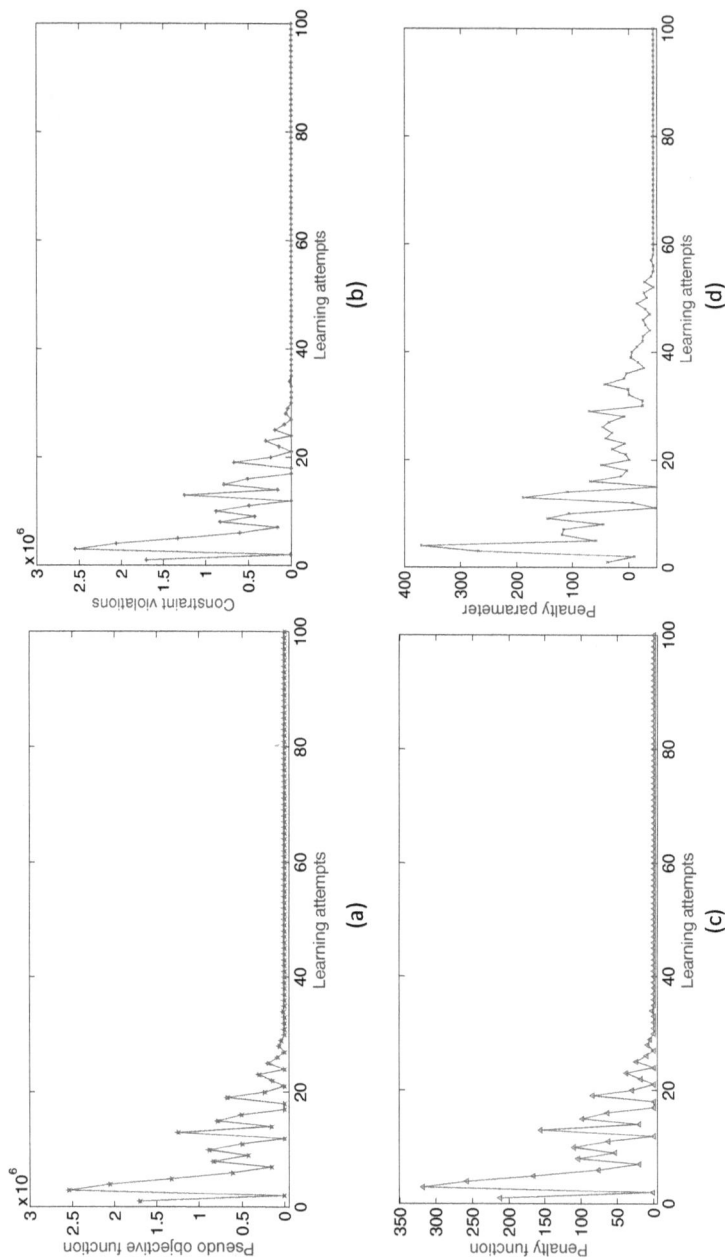

FIGURE 4.4 Behaviour of Pseudo Objective Function (a), Constraint Violations (b), Penalty Function (c) and Penalty Parameter (d) Solving Nonlinear Rosen–Suzuki Convex Programming Problem Using CI-SAPF.

it was observed that the CI solution is dependent on the sampling space reduction factor R which needs to be tuned for every different problem. It necessitates a significant number of preliminary trials. To overcome this limitation, the CI-SAPF can be hybridized with the Colliding Bodies Optimisation (CBO); this is discussed in Chapter 5.

REFERENCES

Coello, C.A.C. (2000) 'Use of a self-adaptive penalty approach for engineering optimization problems', *Computer in Industry*, Vol. 41, pp. 113–127.

He, Q. and Wang, L. (2007) 'An effective co-evolutionary particle swarm optimization for constrained engineering design problem', *Engineering Applications of Artificial Intelligence*, Vol. 20, No. 1, pp. 89–99.

Kale, I.R. and Kulkarni, A.J. (2018) 'Cohort intelligence algorithm for discrete and mixed variable engineering problems', *International Journal of Parallel, Emergent and Distributed Systems*, Vol. 33, No. 6, pp. 627–662.

Kannan, B.K. and Kramer, S.N. (1994) 'An augmented Lagrange multiplier based method for mixed integer discrete continuous optimization and its applications to mechanical design', *Journal of Mechanical Design*, Vol. 116, No. 2, pp. 405–411.

Kashan, A.H. (2011) 'An efficient algorithm for constrained global optimization and application to mechanical engineering design: League championship algorithm (LCA)', *Computer-Aided Design*, Vol. 43, pp. 1769–1792.

Kashan A.H. (2015) 'A new metaheuristic for optimization: Optics inspired optimization (OIO)', *Computers and Operations Research*, Vol. 55, pp. 99–125.

Nanakorn, P. and Meesomklin, K. (2001) 'An adaptive penalty function in genetic algorithms for structural design optimization', *Computer and Structures*, Vol. 79, pp. 2527–2539.

Sandgren, E. (1990) 'Nonlinear integer and discrete programming in mechanical design optimization', *Journal of Mechanical Design*, Vol. 112, No. 2, pp. 223–229.

Srivastava, V.K. and Fahim, A. (2001) 'A two-phase optimization procedure for integer programming problems', *Computers and Mathematics with Applications*, Vol. 42, pp. 1585–1595.

CHAPTER 5

Hybridization of Cohort Intelligence with Colliding Bodies Optimisation

5.1 CHARACTERISTICS OF CI

The CI algorithm has five main characteristics, as follows:

1. The socio-inspired CI algorithm is a group of learning candidates. Every candidate has inherently common goal to achieve the best behaviour by improving their qualities by following, interacting, cooperating and competing with itself and other candidates in a cohort to improve their qualities. Eventually, the behaviour of the entire cohort improves.

2. Every candidate observes itself and every other candidate in the cohort in order to improve its individual behaviour and associated qualities.

3. CI is incorporated with a roulette wheel approach which provides the possible choices available to follow the best candidate.

4. In the algorithm for CI, at the end of every learning attempt the cohort candidates independently update their search space using a sampling space reduction factor.

DOI: 10.1201/9781003245193-5

5. The problems when there are a large number of variables can be efficiently handled (Kulkarni et al., 2016c).

5.2 COLLIDING BODIES OPTIMIZATION (CBO)

The CBO algorithm was proposed by Kaveh and Mahdavi (2014). It was motivated by the physical behaviour of colliding bodies (objects). It obeys the law of conservation of momentum and energy in which the momentum of all the objects before collision is equal to the momentum of all the objects after collision. After the collision, two moving bodies having masses and velocities are separated with updated velocities. This causes to move an object towards a better position in the search space. The characteristics of the CBO algorithm are as follows:

1. The CBO algorithm is governed by the law of conservation of momentum and energy. The momentum of all the objects before collision is equal to the momentum of all the objects after collision.

2. As the Colliding Bodies (CBs) are divided into stationary objects and moving objects, it is necessary to consider an even number of CBs for computation.

3. The CBs are arranged in such a way that the moving objects are keen to find better positions in the search space.

4. Moving objects motivate the stationary objects to explore the search space and seek a better solution.

5. The Coefficient of Restitution (COR) is used here to keep the balance between global and local minima.

6. The CBs are independent of computational parameters. So the preliminary process of parameter tuning is not required.

The CBO algorithm was successfully validated on specific continuous, discrete and mixed variable truss structure and design engineering domain optimisation problems (Kaveh and Mahdavi, 2015). It was observed that CBO was sensitive to the number of objects and required more colliding bodies to maintain better convergence and a higher level of exploration. With a smaller number of colliding bodies the solution became trapped close to local optima and was obtained with premature convergence. Furthermore, Kaveh and Mahdavi (2015) modified the CBO to obtain an Enhanced CBO (ECBO) in order to incorporate the colliding memory, thus saving the so far best solutions. In this way, it could be used for further operations by replacing current worst solutions. This results in an enhanced ECBO algorithm that provides faster convergence with less computational cost. In ECBO, in order to make the solution jump out of the local minima two parameters were introduced randomly from the interval [0,1]. The first parameter represented the change in component of each colliding body compared with a second uniformly distributed random number. This further decided the modification in the position of colliding bodies. The hybrid CBO-PSO (Kaveh and Mahdavi, 2015) was proposed to exploit the ability of CBO by incorporating basic features of PSO. The CBO-PSO was successfully validated by solving continuous variable truss structure problems with dynamic constraints incorporated with a SPF approach.

5.3 FRAMEWORK OF CI-SAPF-CBO

It is necessary to generalise the problem-solving technique in order to explore the applicability for diversified real-world applications. CI has already been validated by solving large group of problems; however, it required preliminary trials to set a sampling space reduction factor R to get a better exploration in the search space (Kale and Kulkarni, 2018). In order to overcome this limitation of the CI algorithm, an important characteristic of CBO is incorporated into CI.

The CI-SAPF-CBO algorithm (refer to Figure 5.1) employs CI for global search, SAPF for constraint handling and CBO for local search, and refined the solutions obtained from CI. The natural tendency of CI candidates is to follow candidates chosen probabilistically using a roulette wheel approach (Xu et al., 2013) to evolve individual behaviour. Furthermore, the learning ability of CI candidates is refined (updated) using CBO. The CI-SAPF-CBO is mathematically expressed as follows:

Step 1: Consider a cohort with C candidates; every individual candidate $c(c = 1, 2, \ldots, C)$ has a set of attributes/variables $X^c = \left(x_1^c, x_2^c, \ldots, x_N^c \right)$ that makes up the behaviour of an individual candidate $f(X^c)$. The initial solution is randomly generated, similarly to other population-based techniques, as follows:

$$X^c = \Psi^{lower} + \left(\Psi^{upper} - \Psi^{lower} \right).rand(1, N) \tag{5.1}$$

The round-off integer sampling approach was employed to generate the integer value and aids in selecting the discrete variable from the predefined set.

Step 2: The SAPF approach is incorporated to handle constraints and obtain the pseudo-objective function $\phi(X^c)$ (see Eq. 4.8).

Step 3: The probability of selecting the behaviour $f(X^c)$ of every associated candidate $c(c = 1, 2, ..., C)$ is calculated as follows:

$$p^c = \frac{1/\phi(X^c)}{\sum_{c=1}^{C} 1/\phi(X^c)} \tag{5.2}$$

Step 4: Every individual candidate $c(c = 1, 2, ..., C)$ generates a random number $r \in [0,1]$ and, using a roulette wheel approach, decides to follow the corresponding behaviour $\phi(X^c)$ and associated attributes X^c. The behaviour is selected by candidate c and is not known in advance. The roulette wheel approach provides the opportunity for every behaviour in the cohort to get selected purely based on its quality. In addition, it may also increase the chances of any candidate to select the better behaviour as the associated probability stake p^c, $c(c = 1, 2, ..., C)$ (refer to Eq. (5.2)) in the interval [0,1] is directly proportional to the quality of the behaviour $\phi(X^c)$. In other words, the better the solution, the higher is the probability of being followed by the candidates in the cohort.

Step 5: After following a suitable candidate in the cohort, all the candidates are arranged in descending order of the fitness value. In the context of CBO, the first half of the candidates (those having greater fitness value) are referred to as stationary bodies and the other half of the candidates (having a lesser fitness value) are referred to as moving bodies. For the sake of the hybridization of CI-SAPF-CBO, stationary bodies are considered as slow learning candidates and moving bodies are considered as fast learning candidates. The candidates who learn quickly motivate the slow learning candidates to improve their learning ability in the cohort. The learning ability of every candidate is identified by determining the initial and final velocities of colliding bodies (C candidates) after collision (Kaveh and Mahdavi, 2014). The learning ability of each candidate refers to the velocities (m) of CBs.

The initial learning ability (initial velocity $m^c = 0$) of slow learning candidates and fast learning candidates is represented as follows:

$$m^c = 0, \quad c = 1, 2, \ldots, \frac{C}{2} \tag{5.3}$$

$$m^c = \phi\left(X^{c+\frac{C}{2}}\right) - \phi(X^c), \quad c = \frac{C}{2} + 1, \frac{C}{2} + 2, \ldots, C \tag{5.4}$$

The final learning ability of slow learning candidates and fast learning candidates are represented as follows:

$$m^{lc} = \frac{\left(p^{c+\frac{C}{2}} + \varepsilon p^{c+\frac{C}{2}}\right) m^{c+\frac{C}{2}}}{p^c + p^{c+\frac{C}{2}}}, \quad c = 1, 2, \ldots, \frac{C}{2} \tag{5.5}$$

$$m^{lc} = \frac{\left(p^c + \varepsilon p^{c-\frac{C}{2}}\right) m^c}{p^c + p^{c-\frac{C}{2}}}, \quad c = \frac{C}{2} + 1, \frac{C}{2} + 2, \ldots, C \tag{5.6}$$

By evaluating the initial learning abilities of each of the candidate, the COR is introduced to control the exploration of search space and the exploitation of the best solution (Kaveh and Mahdavi, 2015). More specifically, the COR controls the local and global searches. It is calculated as follows:

$$COR = 1 - \frac{iter}{iter_{max}}$$

where m^c is the initial position, m^{lc} is the final position and $iter$ and $iter_{max}$ are the current iteration number and total number of iterations, respectively. Further, the final learning abilities of fast learning candidates is used to obtain the new position of candidates in the search space, which fulfils the objective of removing the sampling space reduction factor R from CI-SAPF algorithm.

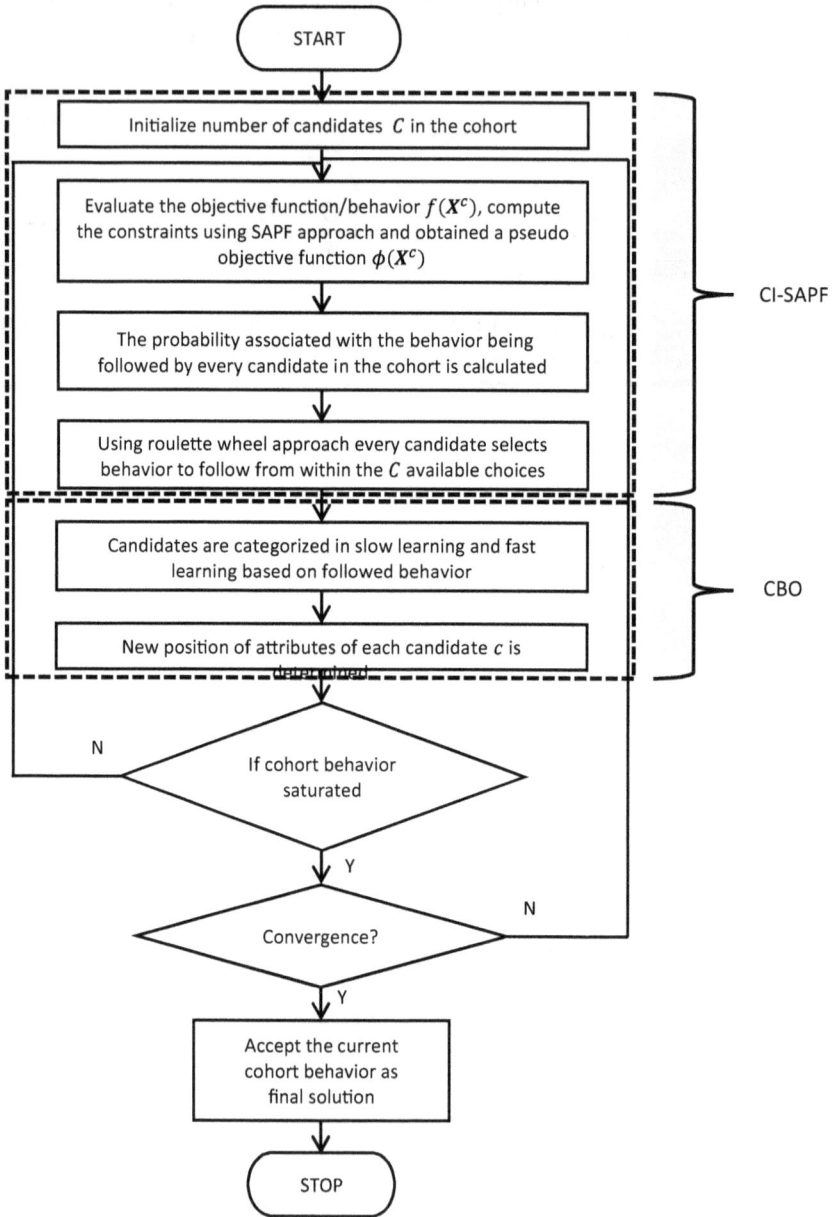

FIGURE 5.1   CI-SAPF-CBO Flowchart.

Step 6: The new positions of attributes for every candidate in the search space are updated as follows:

$$X^{c_{new}} = X^c + rand.m^{lc}, \quad i = 1, 2, \ldots, \frac{C}{2} \tag{5.7}$$

$$X^{c_{new}} = X^{c - \frac{C}{2}} + rand.m^{lc}, \quad c = \frac{C}{2} + 1, \frac{C}{2} + 2, \ldots, C \tag{5.8}$$

where $X^{c_{new}}$, X^c and m^{lc} are the new position of attributes, previous position of attributes and final learning ability of fast learning candidates, respectively, and *rand* is the random vector uniformly distributed in the range [-1,1]. The learning attempt is repeated from Step 2 until the termination criteria (number of iterations) is satisfied.

The MATLAB source code for CBO was adopted from Kaveh and Mahdavi (2015). The CI-SAPF-CB O was coded in MATLAB 7.7.0 (R2013b) and the simulations were run on Windows platform using an Intel(R) Core(TM)2Duo, 2.93GHz processor speed and 4GB RAM. Furthermore, for every individual problem CI-SAPF-CBO was solved iteratively 30 times.

5.4 PROBLEM SOLVED

In this chapter, the same problems are considered as in Chapter 4 in order to validate the performance of the algorithm. The solutions to the rest of the problems are discussed in Chapter 6. The solutions obtained for the 6-bar truss structure, pressure vessel engineering design and linear and nonlinear problems using CI-SAPF-CBO are compared with other contemporary algorithms (refer to Table 5.1, Table 5.2 and Table 5.3).

5.4.1 6-Bar Truss Structure Problem

The CI-SAPF-CBO solution to the 6-bar truss structure problem yielded similar results to that of GA (Nanakorn and Meesomklin, 2001) (see Table 5.1). The best, mean and worst solutions obtained from 30 trials using the CI-SAPF-CBO algorithm are the same (4962.0966 *lb*) with zero standard deviation and where the average computational time was 3.32 *sec*. The average function value reported using GA was 5250 *lb*. From this it is observed that CI-SAPF-CBO performed better than GA. The average number of function evaluations for CI-SAPF-CBO is 2449. This problem is also attempted using CBO. However, it could not yield a feasible solution

TABLE 5.1   Comparison of Results for Solving 6-Bar Truss Structure.

Design Variables (in^2)	GA (Nanakorn and Meesomklin, 2001)	CI-SPF	MRSLS	CI-SAPF	CI-SAPF-CBO
A_1	30	30	30	30	30
A_2	19.9	19.9	19.9	19.9	19.9
A_3	15.5	15.5	15.5	15.5	15.5
A_4	7.22	7.22	7.22	7.22	7.22
A_5	22	22	22	22	22
A_6	22	22	22	22	22
Truss Weight $W(lb)$	4962.1	4962.0966	4962.0966	4962.0966	4962.0966
Function Evaluations	NA	3680	750	2250	1740

due to premature convergence. The convergence curves for CI-SAPF-CBO and CBO are presented in Figure 5.2.

5.4.2 Pressure Vessel Design Engineering Problem

The best, mean and worst solutions obtained for the pressure vessel design problem using CI-SAPF-CBO (see Table 5.2) are 6059.7151, 6066.1112 and 6090.5326, with standard deviation 9.2154, average CPU time 4.29 sec and average function evaluations 12113. From the statistical results it is noted that CI-SAPF obtained better solutions than the other approaches with less computational cost. For the comparison, the CBO was also tested against this problem; however, it could not achieve the feasible solution. The convergence curves for CI-SAPF-CBO and CBO techniques are presented in Figure 5.3. The statistical results for all the solved problems here are presented in Table 6.35.

5.4.3 Linear and Nonlinear Test Problems

The solutions obtained using CI-SAPF-CBO for linear and nonlinear test problems are presented in Table 5.3. For solving the nonlinear Rosen–Suzuki convex programming problem, CI-SAPF-CBO obtained similar results as compared to the two phase optimisation procedure. The CI-SAPF-CBO obtained better results for solving integer linear programming problem for maximisation compared to the two-phase optimisation procedure (Srivastava and Fahim, 2001). The convergence curves for

TABLE 5.2 Comparison of Results for Pressure Vessel Design Problem.

Variables	NIDPM (Sandgren, 1990)	Augmented Lagrange (Kannan and Kramer, 1994)	GA (Coello, 2000)	CPSO (He and Wang, 2007)	LCA (Kashan, 2011)	OIO (Kashan, 2015)	MRSLS	CI-SPF (Kale and Kulkarni, 2018)	CI-SAPF	CI-SAPF-CBO
x_1	1.125	1.125	0.812500	0.812500	NA	NA	0.8125	0.8125	0.75	0.8125
x_2	0.625	0.625	0.437500	0.437500	NA	NA	0.4375	0.4375	0.375	0.4375
x_3	48.97	58.29100	40.32390	42.09126	NA	NA	41.9645	42.0984	40.437	42.0984
x_4	106.72	43.6900	200.0000	176.7465	NA	NA	178.3043	176.6366	198.3782	176.6366
Cost $f(x)$	7981.5690	7198.0428	6288.7445	6061.0777	6059.8553	6059.7143	6076.1215	6059.7152	6051.4819	6059.7152
Function Evaluations	NA	NA	900000	200000	24000	50000	1200	124581	9744	9184

TABLE 5.3   CI-SAPF-CBO Results Solving Linear and Nonlinear Problems.

| Sr. No. | Test Functions | Solver | Search Space | | Function Value | Optimum variables |
			Lower limit	Upper limit		
1	Integer Linear Problem (Maximisation)	Srivastava and Fahim (2001)	[0,…,0]	[200,…,200]	316	[4,87,34,149,0]
		CI-SPF			1037	[200,199,67,104,0]
		MRSLS			1040	[200,200,67,106,0]
		CI-SAPF			1037	[200,199,67,104,0]
		CI-SAPF-CBO			1040	[200,200,67,106,0]
		CBO			393	[111,138,27,0,136]
2	Rosen-Suzuki Convex Programming Problem (Minimisation)	Srivastava and Fahim (2001)	[-10,-10,-10,-10]	[20, 20, 20, 20]	-44	[0,1,2,-1]
		CI-SPF			-44	[0,1,2,-1]
		MRSLS			-44	[0,1,2,-1]
		CI-SAPF			-44	[0,1,2,-1]
		CI-SAPF-CBO			-44	[0,1,2,-1]
		CBO			-32	[-1,1,2,0]

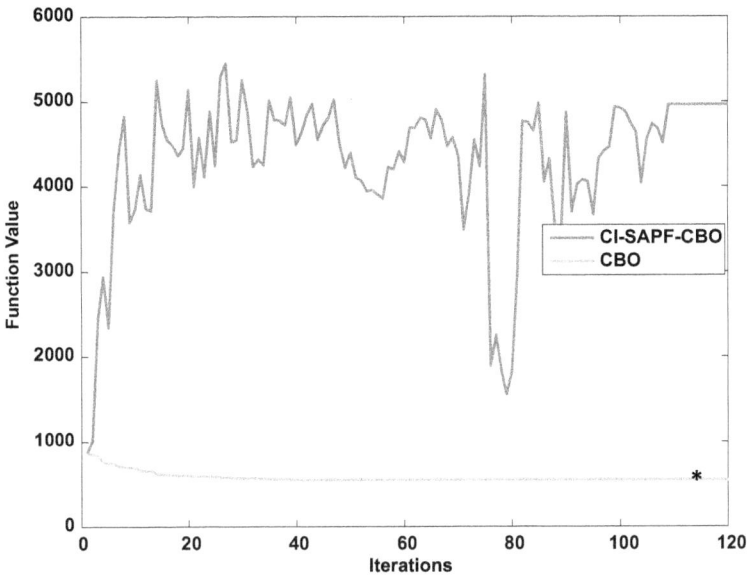

FIGURE 5.2   Comparison of Convergence Curves of CI-SAPF-CBO and CBO for Solving the 6-Bar Truss Problem.

Note: * Infeasible solution.

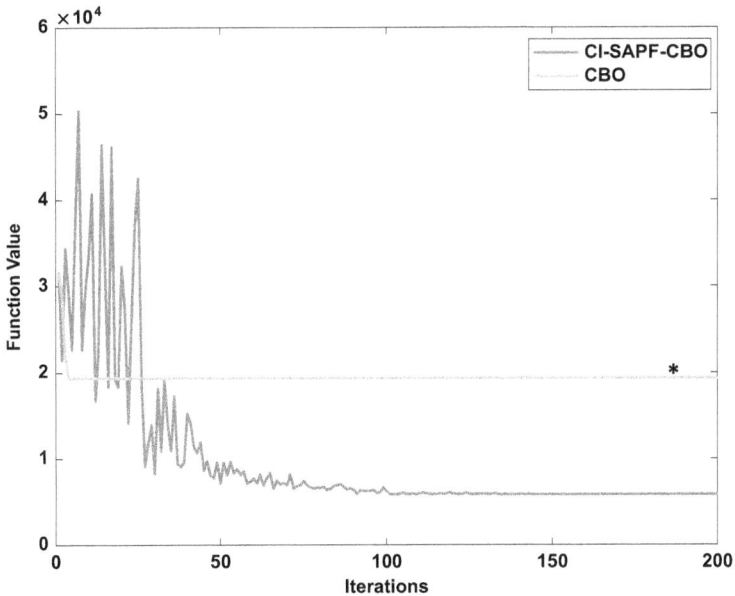

FIGURE 5.3   Comparison of Convergence Curves of CI-SAPF-CBO and CBO for Solving the Pressure Vessel Design Problem.

Note: * Infeasible solution.

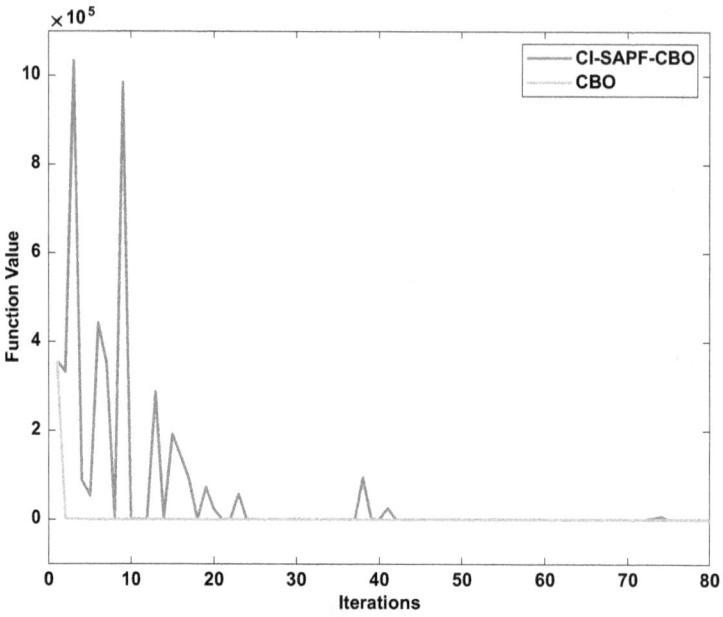

FIGURE 5.4   Comparison of Convergence Curves of CI-SAPF-CBO and CBO for the Rosen–Suzuki Convex Programming Problem.

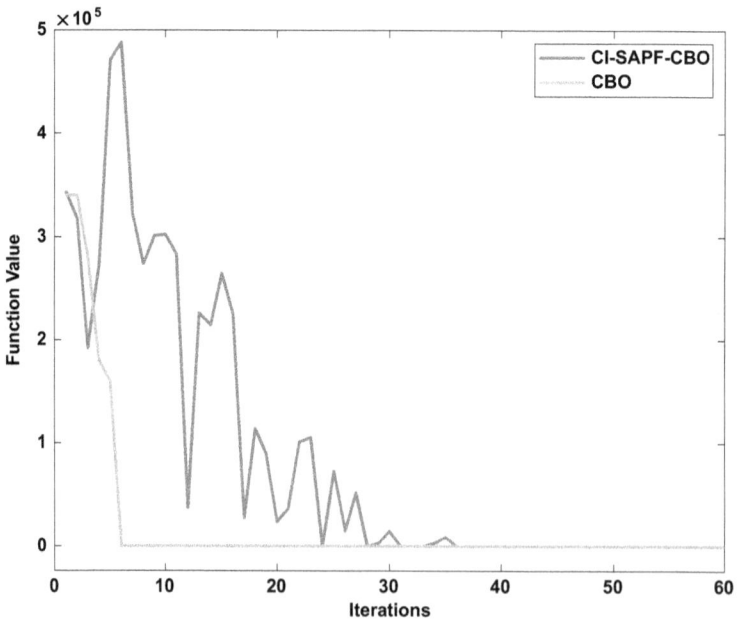

FIGURE 5.5   Comparison of Convergence Curves of CI-SAPF-CBO and CBO for the Integer Linear Programming Problem.

the Rosen–Suzuki convex programming and integer linear problems are illustrated in Figure 5.4 and Figure 5.5, respectively

5.5 CONCLUSION

In this chapter, key characteristics of CI and CBO are combined, resulting in a hybrid CI-SAPF-CBO algorithm. The proposed CI-SAPF-CBO algorithm overcomes the limitation of parameter setting (sampling space reduction factor R) in the CI-SAPF algorithm. To validate the proposed algorithm, the problems from the discrete variable truss structure domain, mixed variable engineering design domain and linear and nonlinear benchmark problems are successfully solved. Compared to the CI-SAPF algorithm, it is observed that the computational cost is significantly reduced. In Chapter 6, the solution to the remaining problems (using CI-SPF, CI-SAPF and CI-SAPF-CBO) are illustrated and thoroughly discussed.

REFERENCES

Coello, C.A.C. (2000) 'Use of a self-adaptive penalty approach for engineering optimization problems', *Computers in Industry*, Vol. 41, pp. 113–127.

He, Q. and Wang, L. (2007) 'An effective co-evolutionary particle swarm optimization for constrained engineering design problem', *Engineering Applications of Artificial Intelligence*, Vol. 20, No. 1, pp. 89–99.

Kale, I.R. and Kulkarni, A.J. (2018) 'Cohort intelligence algorithm for discrete and mixed variable engineering problems', *International Journal of Parallel, Emergent and Distributed Systems*, Vol. 33, No. 6, pp. 627–662.

Kannan, B.K. and Kramer, S.N. (1994) 'An augmented Lagrange multiplier based method for mixed integer discrete continuous optimization and its applications to mechanical design', *Journal of Mechanical Design*, Vol. 116, No. 2, pp. 405–411.

Kashan, A.H. (2011) 'An efficient algorithm for constrained global optimization and application to mechanical engineering design: League championship algorithm (LCA)', *Computer-Aided Design*, Vol. 43, pp. 1769–1792.

Kashan A.H. (2015) 'A new metaheuristic for optimization: Optics inspired optimization (OIO)', *Computers and Operations Research*, Vol. 55, pp. 99–125.

Kaveh, A. and Mahdavi V.R. (2014) 'Colliding bodies optimization: A novel metaheuristic method', *Computers and Structures*, Vol. 39, 15 July, pp. 18–27.

Kaveh, A. and Mahdavi, V.R. (2015) *Colliding Bodies Optimization Extensions and Applications*, Springer.

Kulkarni, A.J., Baki, M.F. and Chaouch, B.A. (2016c) 'Application of the cohort-intelligence optimization method to three selected combinatorial optimization problems', *European Journal of Operational Research*, Vol. 250, No. 2, pp. 427–447.

Nanakorn, P. and Meesomklin, K. (2001) 'An adaptive penalty function in genetic algorithms for structural design optimization', *Computer and Structures*, Vol. 79, pp. 2527–2539.

Sandgren, E. (1990) 'Nonlinear integer and discrete programming in mechanical design optimization', *Journal of Mechanical Design*, Vol. 112, No. 2, pp. 223–229.

Srivastava, V.K. and Fahim, A. (2001) 'A two-phase optimization procedure for integer programming problems', *Computers and Mathematics with Applications*, Vol. 42, pp. 1585–1595.

Xu, W., Geng, Z., Zhu, Q., and Gu, X. (2013) 'A piece wise linear chaotic map and sequential quadratic programming based robust hybrid particle swarm optimization', *Information Sciences*, Vol. 218, pp. 85–102.

CHAPTER 6

Validation of CI-SPF, CI-SAPF and CI-SAPF-CBO for Solving Discrete/ Integer and Mixed Variable Problems

6.1 TRUSS STRUCTURE PROBLEMS

All the truss structure problems aim to solve for the minimisation of weight while satisfying the constraints with maximum allowable stress σ_{max} in both tension and compression of every member, and maximum allowable displacement u_{max} at every node in both the horizontal and vertical directions. In these problems, the number of variables is equal to the number of members of the truss structure. For symmetric truss structure problems such as the 25-bar, 45-bar, 52-bar and 72-bar problems, the variables are considered as a group, as presented in the respective comparison tables. For all truss structure problems, the cross-sectional area is considered to be a variable and it is selected from the set of discrete values. All the truss structure problems are successfully solved by the CI-SPF, CI-SAPF and CI-SAPF-CBO algorithms; however, the CBO algorithm could not achieve the feasible solution for Case 1 and Case 2 of 10-bar and 25-bar truss structure

DOI: 10.1201/9781003245193-6

problems, or 38-bar, 45-bar and 72-bar (Case 2) truss structure problems. This is due to premature convergence of the solution (refer to Figures 6.1–6.8 and 6.11). For the general truss structure problem definition, refer to Eqs. 3.5, 3.6 and 3.7.

6.1.1 10-Bar Truss Structure

The 10-bar truss structure problem of Case 1 and Case 2 is successfully solved using the CI-SPF, CI-SAPF and CI-SAPF-CBO algorithms. The solutions for Case 1 and Case 2 are compared with other contemporary algorithms, as presented in Table 6.1 and Table 6.2, respectively. The best, mean and worst function values (W) obtained for Case 1 and Case 2 using CI-SAPF and CI-SAPF-CBO with standard deviation and the average CPU time is presented in Table 6.3. The other statistical details associated with these problems, along with the parameters used for CI-SPF and CI-SAPF, are listed in Table 6.35. For Case 1, it is noted that CI-SPF obtained the same results as compared to ABC (Sonmez 2011), PC (Kulkarni et al., 2016) and ADS (Hasançebi et al., 2015). CI-SAPF and CI-SAPF-CBO obtained marginally better results as compared to ABC and PC with a smaller number of function evaluations. For solving Case 2, PC yielded a significantly better objective function value with more function evaluations. Compared to ADS, the solution obtained here for Case 1 is similar; however, ADS performance was better in terms of function evaluations. This is due to the Search Dimension Ratio (SDR) incorporated in ADS which helps to explore and exploit the search space and yielded better solutions with fewer function evaluations. As discussed earlier, the CBO could not achieve a feasible solution. The convergence associated with Case 1 and Case 2 of the 10-bar truss structure problem is presented in Figure 6.1 and 6.2, respectively.

6.1.2 Spatial 25-Bar Truss Structure

The 25-bar truss structure problem of Case 1 and Case 2 is successfully solved using CI-SPF, CI-SAPF and CI-SAPF-CBO algorithms. The solutions obtained for Case 1 and Case 2 are compared with other contemporary algorithms presented in Table 6.4 and 6.5, respectively. The statistical results from 30 independent trials are presented in Table 6.6. It is observed that both proposed techniques performed satisfactorily. So far the PC algorithm obtains a better solution with a greater number of function evaluations compared to HS, Discrete Heuristic Particle Swarm Ant Colony

TABLE 6.1 Comparison of Results for Solving the 10-Bar Case 1 Truss Structure Problem.

Design Variables (in^2)	GA (Rajeev and Krishnamoorthy, 1992)	PSO (Li et al., 2009)	PSOPC (Li et al., 2009)	HPSO (Li et al., 2009)	ABC (Sonmez, 2011)	PC (Kulkarni et al., 2016)	ADS (Hasançebi et al., 2015)	MRSLS	CI-SPF (Kale and Kulkarni, 2018)	CI-SAPF	CI-SAPF-CBO
A_1	33.50	30.00	30.00	30.00	33.50	33.50	33.50	30.00	33.50	33.5	33.5
A_2	1.62	1.62	1.62	1.62	1.62	1.62	1.62	1.62	1.62	1.62	1.62
A_3	22	30	26.50	22.90	22.90	22.90	22.90	22.90	22.90	22.9	22.9
A_4	15.50	13.50	15.50	13.50	14.20	14.20	14.20	16.00	14.20	13.9	13.9
A_5	1.62	1.62	1.62	1.62	1.62	1.62	1.62	1.62	1.62	1.62	1.62
A_6	1.62	1.80	1.62	1.62	1.62	1.62	1.62	1.80	1.62	1.62	1.62
A_7	14.20	11.50	11.50	7.97	7.97	7.97	7.97	7.97	7.97	7.97	7.97
A_8	19.90	18.80	18.80	26.50	22.90	22.90	22.90	22.90	22.90	22.9	22.9
A_9	19.90	22.00	22.00	22.00	22.00	22.00	22.00	22.90	22.00	22	22
A_{10}	2.62	1.80	3.09	1.80	1.62	1.62	1.62	1.62	1.62	1.62	1.62
Truss	5613.84	5581.76	5593.44	5531.98	5490.74	5490.74	5490.74	5503.3900	5490.7378	5490.6020	5490.6020
Weight W (lb) Function Evaluations	NA	50000	50000	50000	25800	1852059	1000	2400	19250	16940	14160

TABLE 6.2 Comparison of Results for Solving the 10-Bar Case 2 Truss Structure Problem.

Design Variables (in^2)	PSO (Li et al., 2009)	PSOPC (Li et al., 2009)	HPSO (Li et al., 2009)	MBA (Sadollah et al., 2012)	PC (Kulkarni et al., 2016)	MRSLS	CI-SPF	CI-SAPF	CI-SAPF-CBO
A_1	24.50	31.50	31.50	29.50	23.50	30.5	31	31	31
A_2	0.10	0.10	0.10	0.01	0.10	0.1	0.1	0.1	0.1
A_3	22.50	23.50	24.50	24.00	26.00	23.5	23	23	23
A_4	15.50	18.50	15.50	15.00	14.00	15	15	15	15
A_5	0.10	0.10	0.50	0.01	0.10	0.1	0.1	0.1	0.1
A_6	1.50	0.50	0.50	0.05	2.00	0.5	0.5	0.5	0.5
A_7	8.50	7.50	7.50	7.50	12.50	7.5	7.5	7.5	7.5
A_8	21.50	21.50	20.50	21.50	13.00	21.5	21	21	21
A_9	27.50	23.50	20.50	21.50	20.00	21	21.5	21.5	21.5
A_{10}	0.10	0.10	0.10	0.10	0.10	0.1	0.1	0.1	0.1
Truss Weight $W(lb)$	5243.71	5133.16	5073.51	5067.33	4686.7729	5062.9882	5062.1097	5061.7597	5061.7597
Function Evaluations	NA	NA	NA	NA	2363380	2863	11130	9450	8400

TABLE 6.3   Statistical Results for the 10-Bar Case 1 and 2 Using CI-SPF, CI-SAPF and CI-SAPF-CBO.

Results	CI-SPF		CI-SAPF		CI-SAPF-CBO	
	Case 1	Case 2	Case 1	Case 2	Case 1	Case 2
Best	5490.7387	5062.1097	5490.6021	5061.7696	5490.6021	5061.7696
Mean	5501.1295	5064.7229	5505.7535	5062.2156	5505.7142	5061.8071
Worst	5508.3824	5067.3314	5534.9591	5067.3314	5532.8502	5062.1155
Std. Dev.	12.3321	2.5353	16.6545	1.5402	14.0915	0.1252
Avg. CPU Time *sec.*	12.95	32.56	10.64	12.70	10.80	33.42
Avg. Fun. Eval.	15813	23383	22654	22874	16845	24325

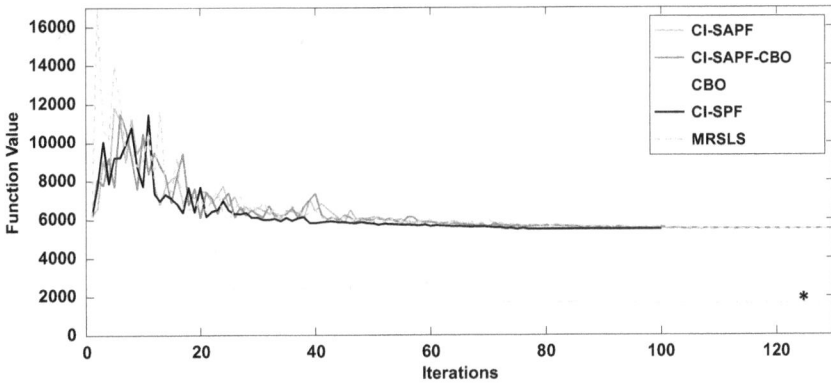

FIGURE 6.1   Comparison of the Convergence Curves of CI-SPF, MRSLS, CI-SAPF, CI-SAPF-CBO and CBO for Solving the 10-Bar Case 1 Truss Problem.

Note: * Infeasible solution.

Optimisation (DHPSACO), PSO, PSOPC and HPSO (Li et al., 2009) a significant improvement is observed in the function value obtained using the proposed algorithms. The other statistical details for CI-SPF, CI-SAPF and CI-SAPF-CBO are presented in Table 6.35. The convergence curves for proposed techniques and CBO are presented in Figures 6.3 and 6.4.

6.1.3 Planer 38-Bar Truss Structure

The 38-bar truss structure problem is successfully solved using CI-SPF, CI-SAPF and CI-SAPF-CBO, and compared with other contemporary techniques as presented in Table 6.7. From the comparison table it is noted that as the algorithm evolved from CI-SPF to CI-SAPF-CBO, the function value also improved. For CI-SPF, CI-SAPF and CI-SAPF-CBO

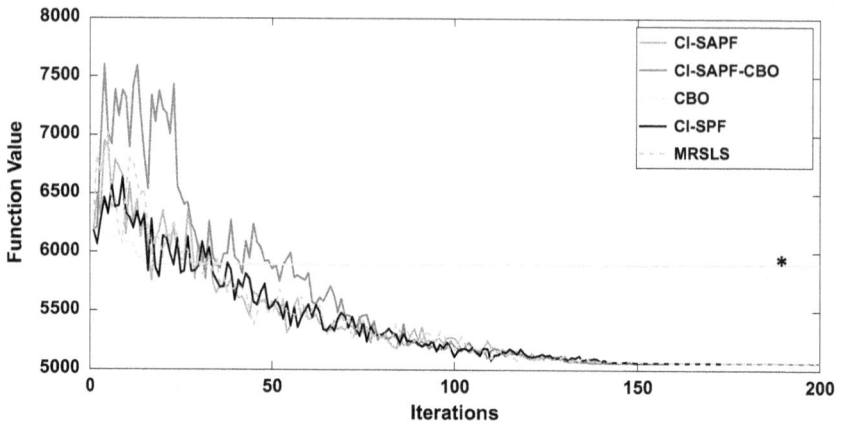

FIGURE 6.2 Comparison of the Convergence Curves of CI-SPF, MRSLS, CI-SAPF, CI-SAPF-CBO and CBO for Solving the 10-Bar Case 2 Truss Problem.

Note: * Infeasible solution.

the best, mean and worst function values (W) along with the standard deviation, average CPU time and average function evaluations are presented in Table 6.8, and the other statistical results are listed in Table 6.35. The solutions reported by the CI-SAPF and CI-SAPF-CBO algorithms are better than that of the PC algorithm. However, the PC algorithm was more robust, having a standard deviation of 0.3696. The convergence curves for proposed techniques and CBO are presented in Figure 6.5.

6.1.4 Planer 45-Bar Truss Structure

For the 45-bar truss structure problem the results obtained using CI-SPF, CI-SAPF and CI-SAPF-CBO algorithms are presented in Table 6.9. The best, mean and worst function values (W) obtained from 30 trials along with the standard deviation, average CPU time and average function evaluations are presented in Table 6.10. The other computational details and parameters associated with the proposed algorithms are listed in Table 6.35. The CI-SAPF and CI-SAPF-CBO algorithms obtained better results compared to other algorithms; however, CI-SAPF-CBO has a lower computational cost (in terms of average function evolutions and average CPU time). The convergence curves for the proposed techniques and CBO are presented in Figure 6.6.

TABLE 6.4 Comparison of Results for Solving the 25-Bar Case 1 Truss Structure Problem.

Design Variables (in^2)	HS (Lee et al., 2005)	DHPSACO (Kaveh and Talatahari, 2009)	PSO (Li et al., 2009)	PSOPC (Li et al., 2009)	HPSO (Li et al., 2009)	PC (Kulkarni et al., 2013)	MRSLS	CI-SPF (Kale and Kulkarni, 2018)	CI-SAPF	CI-SAPF-CBO
A_1	0.01	0.01	0.01	0.01	0.01	0.01	0.4	1.3	0.01	0.01
$A_2 \sim A_5$	2.0	1.6	2.6	2.0	2.0	0.4	4.4	1.3	2	2
$A_6 \sim A_9$	3.6	3.2	3.6	3.6	3.6	3.6	2.4	3.4	2	2
$A_{10} \sim A_{11}$	0.01	0.01	0.01	0.01	0.01	0.01	0.4	0.2	0.4	0.4
$A_{12} \sim A_{13}$	0.01	0.01	0.4	0.01	0.01	2	0.4	2.4	3.2	3.2
$A_{14} \sim A_{17}$	0.8	0.8	0.8	0.8	0.8	0.8	0.01	0.9	0.01	0.01
$A_{18} \sim A_{21}$	1.6	2.0	1.6	1.6	1.6	0.01	2.4	0.1	0.4	0.4
$A_{22} \sim A_{25}$	2.4	2.4	2.4	2.4	2.4	4	1.6	3.4	1.6	1.6
Truss Weight $W(lb)$	560.59	551.61	566.44	560.59	560.59	477.1668	616.2611	516.2024	512.8140	512.8140
Function Evaluations	NA	NA	NA	NA	NA	1844457	3650	21350	18500	15000

TABLE 6.5 Comparison of Results for Solving the 25-Bar Case 2 Truss Structure Problem.

Design Variables (in²)	PSO (Li et al., 2009)	PSOPC (Li et al., 2009)	HPSO (Li et al., 2009)	DHPSACO (Kaveh and Talatahari, 2009)	PC (Kulkarni et al., 2013)	MRSLS	CI-SPF (Kale and Kulkarni, 2018)	CI-SAPF	CI-SAPF-CBO
A_1	1.000	0.111	0.111	0.111	0.111	0.307	0.196	0.111	0.111
$A_2 \sim A_5$	2.620	1.563	2.130	2.130	0.563	0.563	0.141	0.25	0.307
$A_6 \sim A_9$	2.620	3.380	3.380	3.380	3.13	3.47	3.09	3.88	3.88
$A_{10} \sim A_{11}$	0.250	0.111	0.111	0.111	0.141	0.141	0.196	0.141	0.141
$A_{12} \sim A_{13}$	0.307	0.111	0.111	0.111	1.8	2.62	4.18	2.63	2.63
$A_{14} \sim A_{17}$	0.602	0.766	0.766	0.766	0.766	0.766	1.228	0.766	0.766
$A_{18} \sim A_{21}$	1.457	1.990	1.620	1.620	0.111	0.111	0.391	0.25	0.25
$A_{22} \sim A_{25}$	2.880	2.380	2.620	2.620	3.88	3.87	3.55	3.38	3.38
Truss Weight $W(lb)$	567.49	567.49	551.14	551.14	464.1471	491.9080	500.0447	470.1388	473.4724
Function Evaluations	NA	NA	NA	NA	1963415	4050	14350	31500	15000

TABLE 6.6 Statistical Results for the 25-Bar Case 1 and 2 Using CI-SPF, CI-SAPF and CI-SAPF-CBO.

	CI-SPF		CI-SAPF		CI-SAPF-CBO	
Results	Case 1	Case 2	Case 1	Case 2	Case 1	Case 2
Best	516.2024	500.0447	512.8140	470.1388	512.8140	473.4724
Mean	545.3047	521.1982	530.87605	486.79338	531.9475	485.9673
Worst	557.6885	546.1536	548.61476	498.0591	543.0734	487.9375
Std. Dev.	13.1198	12.7870	10.1639	6.8008	8.8008	2.8152
Avg. CPU Time *sec.*	17.65	14.22	12.21	15.95	11.71	22.18
Avg. Fun. Eval.	29711	23002	22588	26796	21652	28513

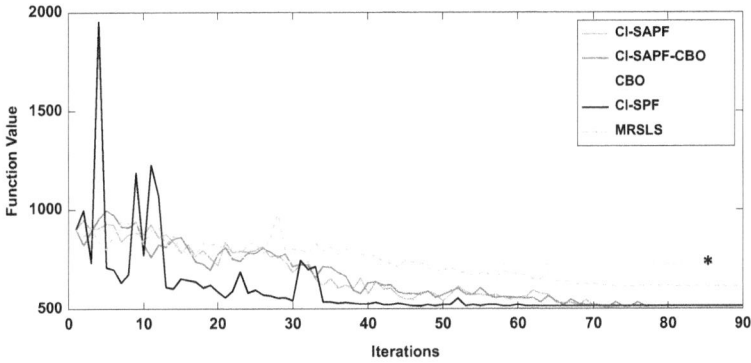

FIGURE 6.3 Comparison of the Convergence Curves of CI-SPF, MRSLS, CI-SAPF, CI-SAPF-CBO and CBO for Solving the 25-Bar Case 1 Truss Problem.

Note: * Infeasible solution.

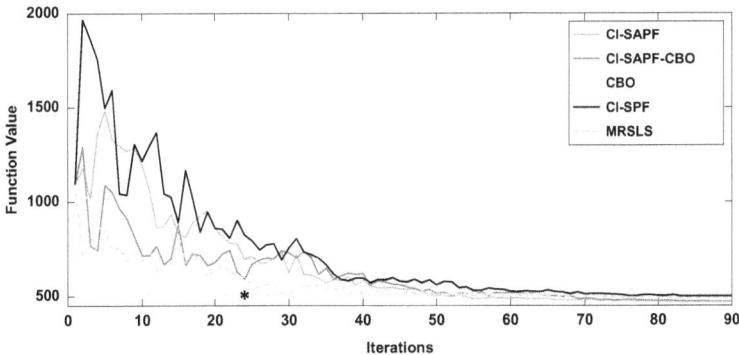

FIGURE 6.4 Comparison of the Convergence Curves of CI-SPF, MRSLS, CI-SAPF, CI-SAPF-CBO and CBO for Solving the 25-Bar Case 2 Truss Problem.

Note: * Infeasible solution.

TABLE 6.7   Comparison of Results for Solving the 38-Bar Truss Structure Problem.

Design Variables (in^2)	ISCSO (Rudolph and Schmidt, 2012)	PC (Kulkarni et al., 2016)	MRSLS	CI-SPF (Kale and Kulkarni, 2018)	CI-SAPF	CI-SAPF-CBO
A_1	14.6	14.6	14.6	14	14.2	13.7
A_2	12.9	12.9	13.9	12.7	12.4	12.1
A_3	11.3	11.5	11.7	10.7	10.3	10.4
A_4	9.7	9.6	10.3	9.6	9	9.5
A_5	8.2	7.8	7.9	8.2	7.5	7.6
A_6	6.5	6.5	6.3	6.6	6.4	6.1
A_7	4.9	4.9	4.6	4.7	4.6	4.5
A_8	3.3	3.1	3.2	3.1	3.1	3
A_9	1.7	1.6	1.6	1.6	1.5	1.5
A_{10}	15	15	15	14.9	15	14.9
A_{11}	14.6	14.8	13.6	14.6	14.1	13.1
A_{12}	12.9	12.8	12.9	13.6	12.4	12.1
A_{13}	11.3	11.3	10.7	11.4	10.7	10.1
A_{14}	9.7	9.7	10.4	9.9	9	9.1
A_{15}	8.2	8.1	8.2	8.5	7.8	7.5
A_{16}	6.5	6.3	6.2	6.5	6.3	6
A_{17}	4.9	4.7	4.7	5.1	4.5	4.3
A_{18}	3.3	3.5	3.6	3.3	3.1	3.1
A_{19}	1.7	1.6	1.5	1.7	1.5	1.5
Truss Weight W(lb)						
Function Evaluations						

TABLE 6.8   Statistical Results for the 38-Bar Truss Structure Problem Using CI-SPF, CI-SAPF and CI-SAPF-CBO.

Results	CI-SPF	CI-SAPF	CI-SAPF-CBO
Best	5900.3465	5891.0472	5889.9511
Mean	5916.9868	5895.3762	5903.2514
Worst	5938.0018	5898.4401	5927.0836
Std. Dev.	10.5244	2.1873	10.9272
Avg. CPU Time *sec.*	42.30	87.80	33.30
Avg. Fun. Eval.	89802	69920	46900

Design Variables (in^2)	ISCSO (Rudolph and Schmidt, 2012)	PC (Kulkarni et al., 2016)	MRSLS	CI-SPF (Kale and Kulkarni, 2018)	CI-SAPF	CI-SAPF-CBO
A_{20}	1.6	1.8	1.6	1.7	1.4	1.5
A_{21}	1.7	1.7	1.8	1.6	1.5	1.5
A_{22}	1.7	2.3	1.7	2.3	1.4	1.5
A_{23}	2.3	2.4	1.6	2.2	1.6	1.5
A_{24}	2.1	2.5	1.7	2.1	1.6	1.5
A_{25}	1.6	1.8	1.7	1.7	1.5	1.5
A_{26}	1.7	1.7	1.6	1.6	1.5	1.5
A_{27}	1.7	2.3	1.6	2.3	1.5	1.5
A_{28}	2.3	2.4	1.7	2.2	1.5	1.4
A_{29}	2.1	2.5	2.3	2.1	2.2	2.1
A_{30}	1.6	1.8	2.3	1.7	2.1	2.2
A_{31}	1.7	1.7	2.3	1.6	2.2	2.1
A_{32}	1.7	2.3	2.3	2.3	2.2	2.1
A_{33}	2.3	2.4	2.4	2.2	2.2	2.2
A_{34}	2.1	2.5	2.2	2.1	2.2	2.2
A_{35}	1.6	1.8	2.5	1.7	2.3	2.1
A_{36}	1.7	1.7	2.1	1.6	2.2	2.2
A_{37}	1.7	2.3	2.5	2.3	2.3	2.1
A_{38}	2.3	2.4	2.1	2.2	2.2	2.2
	5889.9	5893.02273	5904.11	5900.34	5891.0472	5889.9511
	4618	1793966	20900	106666	151240	42750

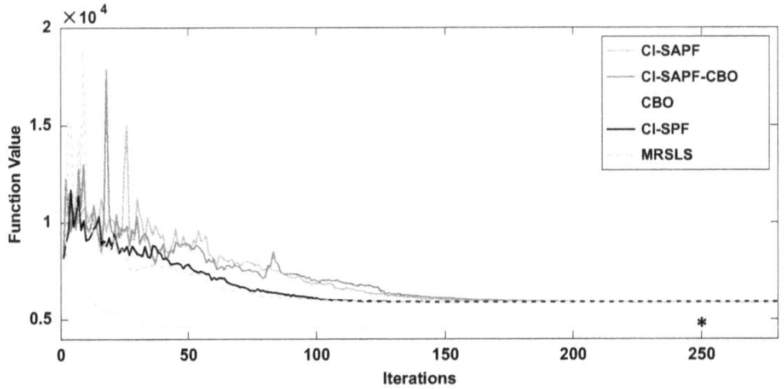

FIGURE 6.5 Comparison of the Convergence Curves of CI-SPF, MRSLS, CI-SAPF, CI-SAPF-CBO and CBO for Solving the 38-Bar Truss Problem.

Note: * Infeasible solution.

TABLE 6.9   Comparison of Results for Solving the 45-Bar Truss Structure Problem.

Design Variables (in^2)	ISCSO (Arnout, 2011)	PC (Kulkarni et al., 2013)	MRSLS	CI-SPF (Kale and Kulkarni, 2018)	CI-SAPF	CI-SAPF-CBO
A_1, A_{44}	9.1	9.7	8.5	8.5	9.1	9.1
A_2, A_{45}	6.8	7	7.80	6.9	7.8	7.8
A_3, A_{43}	4.6	5	5.4	5.6	4.5	4.5
A_4, A_{39}	8.2	7.9	8.7	6.6	8.1	8.1
A_5, A_{41}	2.5	2.3	1.8	0.9	2.6	2.6
A_6, A_{40}	5.2	5.5	5.5	6.6	5.4	5.4
A_7, A_{42}	3.1	3.3	3.3	2.3	3	3
A_8, A_{38}	0.1	0.3	0.2	1.8	0.1	0.1
A_9, A_{34}	15	14.9	14.5	14.2	14.9	14.9
A_{10}, A_{36}	5.1	4.8	5.3	4	4.8	4.8
A_{11}, A_{35}	1.7	1.8	1.8	2.4	1.8	1.8
A_{12}, A_{37}	0.1	0.1	0.1	0.1	0.1	0.1
Truss Weight $W(lb)$						
Function Evaluations						

TABLE 6.10   Statistical Results for the 45-Bar Truss Structure Problem Using CI-SPF, CI-SAPF and CI-SAPF-CBO.

Results	CI-SPF	CI-SAPF	CI-SAPF-CBO
Best	14354.2710	14322.2890	14322.2890
Mean	14390.4402	14476.4990	14413.8947
Worst	14410.2121	14667.1001	14480.4395
Std. Dev.	16.7597	91.7773	40.1861
Avg. CPU Time *sec.*	80.45	64.69	56.95
Avg. Fun. Eval.	131378	114525	103237

6.1.5 Spatial 52-Bar Truss Structure

The discrete variable 52-bar spatial truss structure problem for weight minimisation has been solved successfully using the CI-SPF, CI-SAPF and CI-SAPF-CBO algorithms; the solutions are compared with other contemporary approaches as presented in Table 6.11. The best, mean and worst solutions obtained from 30 trials using proposed techniques with standard deviation, average CPU time and average number of function evaluations

Design Variables (in^2)	ISCSO (Arnout, 2011)	PC (Kulkarni et al., 2012)	MRSLS	CI-SPF (Kale and Kulkarni, 2018)	CI-SAPF	CI-SAPF-CBO
A_{13}, A_{33}	0.1	0.1	0.1	0.1	0.1	0.1
A_{14}, A_{29}	15	15	15	15	15	15
A_{15}, A_{31}	1.8	1.9	1.8	2.4	1.9	1.9
A_{16}, A_{30}	3.2	2.9	3.5	2.4	3	3
A_{17}, A_{32}	6	5.6	5.4	5	5.8	5.8
A_{18}, A_{28}	0.1	0.1	0.1	0.1	0.1	0.1
A_{19}, A_{24}	15	15	15	15	15	15
A_{20}, A_{26}	2.9	3	2.6	2.2	3.1	3.1
A_{21}, A_{25}	0.1	0.1	0.1	0.4	0.1	0.1
A_{22}, A_{27}	7.6	7.1	7.5	7.2	6.8	6.8
A_{23}	0.6	0.8	0.5	0.1	0.5	0.5
	14341.21	14377.92	14420.85	14354.27	14322.2890	14322.2890
	NA	4494278	17100	163483	95625	51750

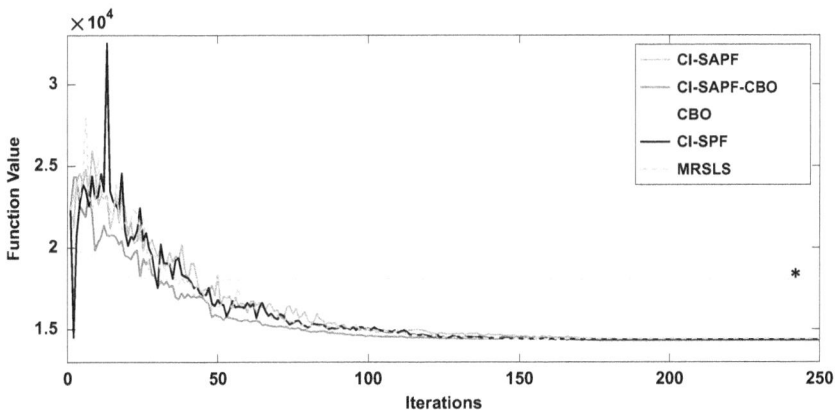

FIGURE 6.6 Comparison of the Convergence Curves of CI-SPF, MRSLS, CI-SAPF, CI-SAPF-CBO and CBO for Solving the 45-Bar Truss Problem.

are presented in Table 6.12. The other associated statistical details are listed in Table 6.35. The reported solution using CI-SAPF-CBO is better compared to other contemporary approaches. The convergence curves for the proposed techniques are presented in Figure 6.7.

TABLE 6.11   Comparison of Results for Solving the 52-Bar Truss Structure Problem.

Design Variables (mm^2)	GA (Wu and Chow, 1995)	HS (Lee et al., 2005)	HPSO (Li et al., 2009)	DHPSACO (Kaveh and Talatahari, 2009)
$A_1 \sim A_4$	4658.055	4658. 55	4658.055	4658.055
$A_5 \sim A_{10}$	1161.288	1161.288	1161.288	1161.288
$A_{11} \sim A_{13}$	645.160	506.451	363.225	494.193
$A_{14} \sim A_{17}$	3303.219	3303.219	3303.219	3303.219
$A_{18} \sim A_{23}$	1045.159	940.000	940.000	1008.385
$A_{24} \sim A_{26}$	494.193	494.193	494.193	285.161
$A_{27} \sim A_{30}$	2477.414	2290.318	2238.705	2290.318
$A_{31} \sim A_{36}$	1045.159	1008.385	1008.385	1008.385
$A_{37} \sim A_{39}$	285.161	2290.318	388.386	388.386
$A_{40} \sim A_{43}$	1696.771	1535.481	1283.868	1283.868
$A_{44} \sim A_{49}$	1045.159	1045.159	1161.288	1161.288
$A_{50} \sim A_{52}$	641.289	506.451	729.256	506.451
Weight (kg)	1970.142	1906.76	1905.495	1904.83
Function Evaluations	60000	NA	100000	5300

TABLE 6.12   Statistical Results for the 52-Bar Truss Structure Problem Using CI-SPF, CI-SAPF and CI-SAPF-CBO.

Results	CI-SPF	CI-SAPF	CI-SAPF-CBO
Best	1894.4817	1894.4817	1891.4391
Mean	1929.7643	1913.9750	1909.3269
Worst	1990.3652	1934.9403	1920.1055
Std. Dev.	27.0146	11.9890	7.1767
Avg. CPU Time *sec.*	71.33	52.73	61.17
Avg. Fun. Eval.	124581	101751	95645

6.1.6 Spatial 72-Bar Truss Structure

The CI-SPF, CI-SAPF and CI-SAPF-CBO algorithms are successfully validated when solving Case 1 and Case 2 of the spatial 72-bar truss structure problem. The solution comparison with other approaches is presented in Table 6.13 and Table 6.14, respectively. So far PC yields a better solution for this problem, while the CI-SAPF and CI-SAPF-CBO algorithms yielded comparable solutions with less computational time and function evaluations. The best, mean and worst function (W) values, standard deviation, average function evaluations and average CPU time obtained in 30

MBA (Sadollah et al., 2012)	CBO (Kaveh and Mahdavi, 2015)	MRSLS	CI-SPF	CI-SAPF	CI-SAPF-CBO
4658.055	4658.055	4658.055	4658.055	4658.055	4658.055
1161.288	1161.288	1161.288	1161.288	1161.288	1161.288
494.193	388.386	494.193	506.451	506.451	363.225
3303.219	3303.219	3303.219	3303.219	3303.219	3303.219
940.000	939.998	939.998	939.998	939.998	939.998
494.193	506.451	494.193	506.451	506.451	494.193
2283.705	2238.705	2238.705	2238.705	2238.705	2238.705
1008.385	1008.385	1008.385	1008.385	1008.385	1008.385
494.193	506.451	506.451	285.161	285.161	388.386
1283.868	1283.868	1283.868	1283.868	1283.868	1283.868
1161.288	1161.288	1161.288	1161.288	1161.288	1161.288
494.193	506.451	494.193	506.451	506.451	494.193
1902.605	1899.35	1903.1809	1894.4817	1894.4817	1891.4391
NA	3840	1000	111584	175240	42432

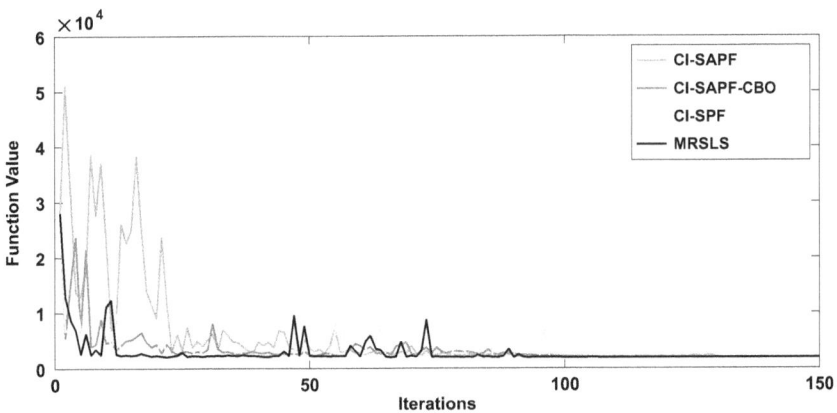

FIGURE 6.7 Comparison of the Convergence Curves of CI-SPF, MRSLS, CI-SAPF, CI-SAPF-CBO and CBO for Solving the 52-Bar Truss Problem.

trials are presented in Table 6.15. The convergence curves for the proposed techniques are presented in Figures 6.8 and 6.9. The other associated results and parameters used to run CI-SPF, CI-SAPF and CI-SAPF-CBO algorithms are illustrated in Table 6.35.

TABLE 6.13 Comparison of Results for Solving the 72-Bar Case 1 Truss Structure Problem.

Design Variables (in^2)	GA (Wu and Chow, 1995)	HPSO (Li et al., 2009)	DHPSACO (Kaveh and Talatahari, 2009)	PC (Kulkarni et al., 2013)	CBO (Kaveh and Mahdavi, 2015)	MRSLS (Kale and Kulkarni, 2018)	CI-SPF (Kale and Kulkarni, 2018)	CI-SAPF	CI-SAPF-CBO
$A_1 \sim A_4$	1.5	2.1	1.9	2.1	1.9	1.6	1.8	2.1	2.1
$A_5 \sim A_{12}$	0.7	0.6	0.5	0.5	0.5	0.7	0.5	0.5	0.5
$A_{13} \sim A_{16}$	0.1	0.1	0.1	0.1	0.1	0.1	0.1	0.1	0.1
$A_{17} \sim A_{18}$	0.1	0.1	0.1	0.1	0.1	0.1	0.1	0.1	0.1
$A_{19} \sim A_{22}$	1.3	1.4	1.3	1.2	1.3	1.3	1.4	1.2	1.2
$A_{23} \sim A_{30}$	0.5	0.5	0.5	0.5	0.5	0.5	0.6	0.5	0.5
$A_{31} \sim A_{34}$	0.2	0.1	0.1	0.1	0.1	0.1	0.1	0.1	0.1
$A_{35} \sim A_{36}$	0.1	0.1	0.1	0.5	0.1	0.1	0.1	0.1	0.1
$A_{37} \sim A_{40}$	0.5	0.5	0.6	0.5	0.6	0.4	0.4	0.5	0.5
$A_{41} \sim A_{48}$	0.5	0.5	0.5	0.1	0.5	0.4	0.5	0.5	0.5
$A_{49} \sim A_{52}$	0.1	0.1	0.1	0.1	0.1	0.1	0.5	0.1	0.1
$A_{53} \sim A_{54}$	0.2	0.1	0.1	0.1	0.1	0.1	0.1	0.2	0.2
$A_{55} \sim A_{58}$	0.2	0.2	0.2	0.5	0.2	0.1	0.1	0.1	0.1
$A_{59} \sim A_{66}$	0.5	0.5	0.6	0.5	0.6	0.6	0.5	0.5	0.5
$A_{67} \sim A_{70}$	0.5	0.3	0.4	0.4	0.4	0.4	0.3	0.4	0.4
$A_{71} \sim A_{72}$	0.7	0.7	0.6	0.6	0.6	0.6	0.6	0.5	0.5
Truss Weight $W(lb)$	400.66	388.94	385.54	372.4095	385.54	381.88	373.5427	372.4095	372.4095
Function Evaluations	NA	NA	5330	8843207	5330	21600	99288	72000	38880

TABLE 6.14 Comparison of Results for Solving the 72-Bar Case 2 Truss Structure Problem.

Design Variables (in^2)	GA (Wu and Chow, 1995)	HPSO (Li et al., 2009)	DHPSACO (Kaveh and Talatahari, 2009)	PC (Kulkarni et al., 2013)	MRSLS (Kale and Kulkarni, 2018)	CI-SPF (Kale and Kulkarni, 2018)	CI-SAPF	CI-SAPF-CBO
$A_1 \sim A_4$	0.196	4.970	1.800	1.8	1.620	2.13	1.62	1.8
$A_5 \sim A_{12}$	0.602	1.228	0.442	0.563	0.563	0.443	0.602	0.443
$A_{13} \sim A_{16}$	0.307	0.111	0.141	0.111	0.111	0.111	0.111	0.111
$A_{17} \sim A_{18}$	0.766	0.111	0.111	0.141	0.111	0.111	0.111	0.111
$A_{19} \sim A_{22}$	0.391	2.880	1.228	1.457	1.228	1.457	1.266	1.266
$A_{23} \sim A_{30}$	0.391	1.457	0.563	0.443	0.602	0.563	0.443	0.563
$A_{31} \sim A_{34}$	0.141	0.141	0.111	0.111	0.111	0.111	0.111	0.111
$A_{35} \sim A_{36}$	0.111	0.111	0.111	0.111	0.111	0.111	0.111	0.111
$A_{37} \sim A_{40}$	1.800	1.563	0.563	0.602	0.602	0.443	0.443	0.391
$A_{41} \sim A_{48}$	0.602	1.228	0.111	0.443	0.443	0.443	0.563	0.563
$A_{49} \sim A_{52}$	0.141	0.111	0.250	0.111	0.111	0.443	0.111	0.111
$A_{53} \sim A_{54}$	0.307	0.196	0.196	0.111	0.111	0.111	0.141	0.141
$A_{55} \sim A_{58}$	1.563	0.391	0.563	0.111	0.111	0.111	0.111	0.111
$A_{59} \sim A_{66}$	0.766	1.457	0.442	0.563	0.563	0.602	0.563	0.563
$A_{67} \sim A_{70}$	0.141	0.766	0.766	0.443	0.307	0.443	0.443	0.443
$A_{71} \sim A_{72}$	0.111	1.563	1.563	0.563	0.602	0.391	0.443	0.443
Truss Weight $W(lb)$	427.203	933.09	393.380	379.9079	380.94	381.3418	380.1807	379.0668
Function Evaluations	NA	NA	NA	8730598	28800	63000	48240	60912

TABLE 6.15   Statistical Results for the 72-Bar Cases 1 and 2 Truss Structure Problem Using CI-CPF, CI-SAPF and CI-SAPF-CBO.

Results	CI-SPF		CI-SAPF		CI-SAPF-CBO	
	Case 1	Case 2	Case 1	Case 2	Case 1	Case 2
Best	373.5426	381.3418	372.4095	380.1807	372.4095	379.0668
Mean	385.6413	385.8208	381.5319	383.2420	398.0869	387.8078
Worst	391.6095	395.1080	384.6292	385.9565	418.9014	397.3978
Std. Dev.	5.1138	3.1475	3.2536	1.6929	11.8925	5.4302
Avg. CPU Time *sec.*	64.36	131.18	78.61	118.52	53.88	80.39
Avg. Fun. Eval.	94012	181496	85500	124080	85872	91872

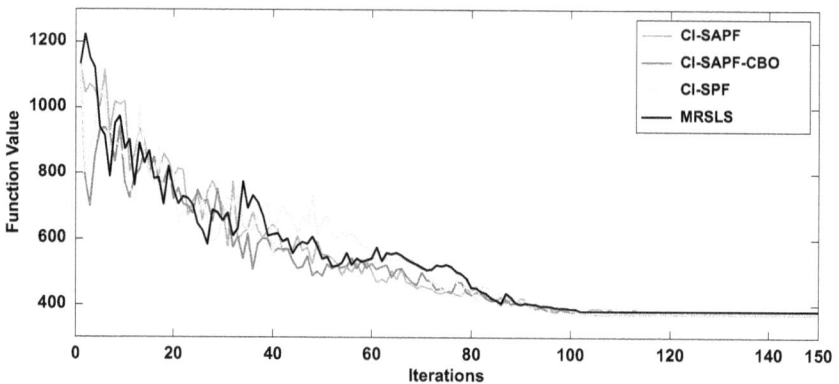

FIGURE 6.8 Comparison of the Convergence Curves of CI-SPF, MRSLS, CI-SAPF, CI-SAPF-CBO and CBO for Solving the 72-Bar Case 1 Truss Problem.

Note: * Infeasible solution.

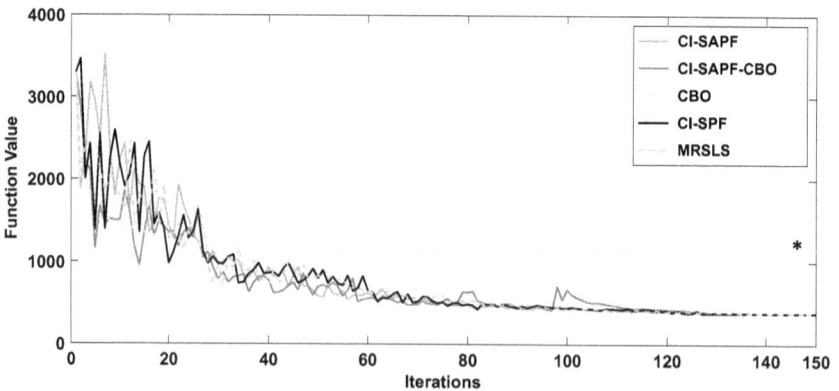

FIGURE 6.9 Comparison of the Convergence Curves of CI-SPF, MRSLS, CI-SAPF, CI-SAPF-CBO and CBO for Solving the 72-Bar Case 2 Truss Problem.

6.2 DESIGN ENGINEERING PROBLEMS

The CI-SPF, CI-SAPF and CI-SAPF-CBO algorithms are tested by solving ten problems from the design engineering domain: stepped cantilever beam problem (minimisation of volume); speed reducer problem (minimisation of weight); reinforced concrete beam problem (minimisation of cost); welded beam design problem Case 1 (minimisation of cost); welded beam design problem Case 2 (minimisation of overall fabrication cost); multiple disc clutch brake problem (minimisation of mass); helical tension compression spring problem (minimisation of volume); I-beam (minimisation of vertical deflection); cantilever beam problem (minimisation of weight); and compound gear problem (minimisation of gear ratio). For the sake of comparison, the CBO algorithm is also used to solve the same problems except for the welded beam Case 2 (formerly reported in CBO (Kaveh and Mahdavi, 2015)). These problems have mixed design variables. For the statistical analysis the CI-SPF, CI-SAPF, CI-SAPF-CBO and CBO are run 30 times for every problem.

6.2.1 Stepped Cantilever Beam Design Problem

The stepped cantilever beam design problem for volume minimisation was proposed by Thanedar and Vanderplaats (1995). Some design variables are discrete $(b_1, h_1, b_2, h_2, b_3, h_3)$ and some continuous (b_4, h_4, b_5, h_5). This problem was solved using Branch & Bound and the SA approach (Thanedar and Vanderplaats 1995). The CI-SPF, CI-SAPF, CI-SAPF-CBO and CBO are also used to solve this problem and the results are compared with other algorithms (see Table 6.16). The best, mean and worst function values along with standard deviation, average function evaluations and average CPU time obtained from applied techniques are presented in Table 6.17. It is observed that CI-SPF, CI-SAPF and CI-SAPF-CBO reported better solutions as compared to other contemporary algorithms, and marginally inferior solutions as compared to FA. The associated statistical results for CI-SPF, CI-SAPF and CI-SAPF-CBO are presented in Table 6.35. The convergence curves for the proposed techniques and CBO are presented in Figure 6.10.

6.2.2 Speed Reducer Design Problem

The solutions to the speed reducer design problem using the CI-SPF, CI-SAPF, CI-SAPF-CBO and CBO algorithms are presented in Table 6.18 along with a comparison with other contemporary approaches. The

TABLE 6.16   Comparison of Results for Solving the Stepped Cantilever Beam Design Problem.

Design Variables	B&B – RU (Thanedar and Vanderplaats 1995)	B&B – CAD (Thanedar and Vanderplaats 1995)	GA – APM (Lemonge and Barbosa 2004)	AIS – GA (Bernardino et al., 2007)	AIS – GA – C (Bernardino et al., 2007)
b_1	4.0000	3.0000	3.0000	3.0000	3.0000
h_1	62.0000	60.0000	60.0000	60.0000	60.0000
b_2	3.1000	3.1000	3.1000	3.1000	3.1000
h_2	60.0000	55.0000	55.0000	55.0000	60.0000
b_3	2.6000	2.6000	2.6000	2.6000	2.6000
h_3	55.0000	50.0000	50.0000	50.0000	50.0000
b_4	2.2050	2.2790	2.2890	2.2350	2.3110
h_4	44.0900	45.5530	45.6260	44.3950	43.1680
b_5	1.7510	1.7500	1.7930	2.0040	2.2250
h_5	35.0300	35.0040	34.5930	32.8790	31.2500
Volume $f(cm^3)$	73555.00	64558.00	64698.56	65559.60	66533.47
Function Evaluations	NA	NA	NA	NA	NA

TABLE 6.17   Statistical Results for the Stepped Cantilever Beam Problem Using CI-SPF, CI-SAPF, CI-SAPF-CBO and CBO.

Results	CI-SPF	CI-SAPF	CI-SAPF-CBO	CBO
Best	63893.4914	63839.4544	63893.4682	80329.3686
Mean	63905.88	64248.9590	64229.2335	85830.5151
Worst	63946.77	64333.4510	64579.4303	101024.2728
Std. Dev.	13.4497	175.2922	207.1399	6610.6072
Avg. CPU Time *sec.*	3.53	5.36	7.46	1.14
Avg. Fun. Eval.	14369	21552	30437	4448

proposed approaches yield better results with less computational cost and function evaluations. The best, mean and worst function values with standard deviation, average computational time and average number of function evaluations are presented in Table 6.18. The other associated parameters are listed in Table 6.35. The CI-SAPF-CBO algorithm yielded the best solution so far with little computational cost (see Table 6.19). The convergence curves for the proposed techniques and CBO are presented in Figure 6.11.

FA (Gandomi et al., 2011)	PC (Kulkarni et al., 2016)	CI-SPF (Kale and Kulkarni, 2018)	MRSLS	CBO	CI-SAPF	CI-SAPF-CBO
3.0000	3.0000	3.0000	3.0000	4.0000	3.0000	3.0000
60.0000	60.0000	60.0000	60.0000	55.0000	60.0000	60.0000
3.1000	3.1000	3.1000	3.1000	3.1000	3.1000	3.1000
55.0000	55.0000	55.0000	55.0000	55.0000	55.0000	55.0000
2.6000	2.6000	2.6000	2.6000	3.1000	2.6000	2.6000
50.0000	50.0000	50.0000	50.0000	52.0000	50.0000	50.0000
2.2050	2.2679	2.2046	2.2100	3.0458	2.2046	2.2045
44.0910	45.3540	44.0911	44.0600	42.3208	44.0910	44.0910
1.7500	3.6500	1.7498	1.7500	2.7314	1.7497	1.7497
34.9950	31.3420	34.9951	34.9900	45.2605	34.9951	34.9951
63893.52	61473.93	63893.49	63903.41	80329.3686	63893.4544	63893.4682
NA	5075700	19740	1460	4560	10000	12000

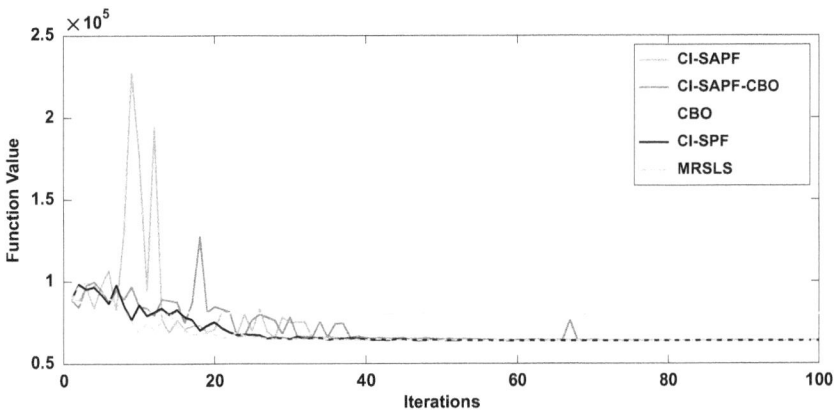

FIGURE 6.10 Comparison of the Convergence Curves of CI-SPF, MRSLS, CI-SAPF, CI-SAPF-CBO and CBO for Solving the Stepped Cantilever Beam Problem.

6.2.3 Reinforced Concrete Beam Design

The solutions to the reinforced concrete beam design problem using the CI-SPF, CI-SAPF and CI-SAPF-CBO approaches reported similar results with less computational cost compared with FA and PC (see Table 6.20).

TABLE 6.18 Comparison of Results for Solving the Speed Reducer Problem for Weight Optimisation.

Design Variables and Constraints	AIS-GA-C (Bernardino et al., 2007)	AIS-GA (Coello and Cortes, 2004)	EA (Efren et al., 2003)	PC (Kulkarni et al., 2016)	CI-SPF (Kale and Kulkarni, 2018)	MRSLS	CBO	CI-SAPF	CI-SAPF-CBO
b	3.500000	3.500000	3.506163	3.506102231	3.500120116	3.5000	3.501114381	3.500007435	3.499999742
z	0.700000	0.700000	0.700831	0.700173506	0.700006522	0.7000	0.700166432	0.700000253	0.70000003
Z	17.0	17.0	17.0	17.0	17.0	17.0	17	17	17
D_1	7.3000035	7.300008	7.460181	7.315691279	7.383509259	8.0000	7.703261947	7.327317877	7.348038872
D_2	7.7153225	7.715322	7.962143	7.778379574	7.423721142	7.4600	7.414969652	7.429919999	7.406565976
L_1	3.3502147	3.350215	3.362900	3.356601485	3.352433103	3.3500	3.424789122	3.35233413	3.352296463
L_2	5.2866545	5.286655	5.308949	5.000832551	5.000013613	5.0000	5.000050534	5.000022599	5.000000468
Weight $W(lb)$	2994.4712	2994.3419	3025.0051	2828.5863	2817.5580	2824.19	2840.4404	2817.0905	2816.7889
Function Evaluations	36000	150000	36000	1132700	16513	1400	1848	5145	8442

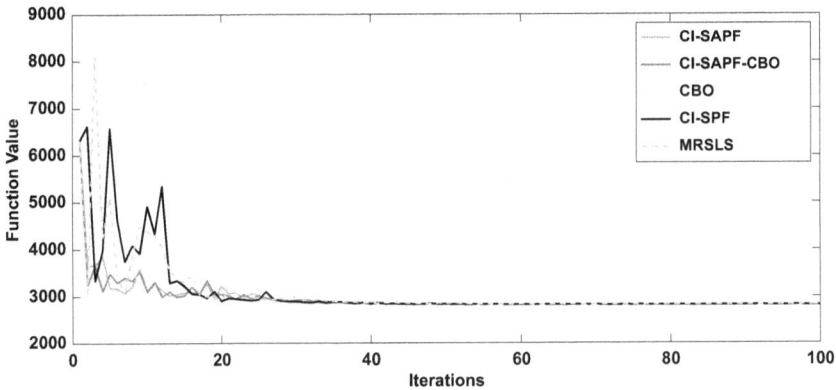

FIGURE 6.11 Comparison of the Convergence Curves of CI-SPF, MRSLS, CI-SAPF, CI-SAPF-CBO and CBO for Solving the Speed Reducer Design Problem.

TABLE 6.19   Statistical Results for the Speed Reducer Design Problem Using CI-SPF, CI-SAPF, CI-SAPF-CBO and CBO.

Results	CI-SPF	CI-SAPF	CI-SAPF-CBO	CBO
Best	2817.5581	2817.0910	2816.3410	2840.4404
Mean	2824.9781	2820.3190	2817.6529	3534.0041
Worst	2828.4442	2825.8710	2819.5744	3582.7451
Std. Dev.	2.7850	2.0660	1.0276	264.3994
Avg. CPU Time *sec.*	2.69	2.25	1.03	0.30
Avg. Fun. Eval.	10851	14420	7597	613

The statistical results for CI-SPF, CI-SAPF, CI-SAPF-CBO and CBO are presented in Table 6.21. The convergence curves for both proposed techniques and CBO are presented in Figure 6.12. The statistical results for all the proposed techniques are presented in Table 6.35.

6.2.4 Welded Beam Design Case 1 and Case 2

The welded beam design problem of Case 1 and Case 2 is successfully solved using the CI-SPF, CI-SAPF and CI-SAPF-CBO algorithms. The solutions are compared with other contemporary techniques. It is observed that the proposed techniques performed better than the GA, PSO and CBO algorithms (see Table 6.22 and Table 6.23). For Case 1, CBO was previously used by Kaveh and Mahdavi (2014), and in the current work CBO is tested on Case 2. For both the cases CI-SAPF and CI-SAPF-CBO performed exceptionally better. The statistical results for Case 1 and Case 2 obtained

TABLE 6.20   Comparison of Results for Solving the Reinforced Concrete Beam Design Problem.

Design Variables and Constraints	FA (Gandomi et al., 2011)	PC (Kulkarni et al., 2016)	CI-SPF (Kale and Kulkarni, 2018)	MRSLS	CBO	CI-SAPF	CI-SAPF-CBO
A_i	6.32	6.32	7.11	6.32	7.11	6.32	6.32
b	34	34	32	34	32	34	34
h	8.50	8.5	8.6297	8.5	8	8.5	8.5
g_1	−0.2241	−0.2241	−0.0002	−0.7745	0	−3.37E-09	−3.27E-07
g_2	0	0	−0.2256	−0.0635	−0.9173	−2.24E-01	−0.2241
Min. Cost f	359.2080	359.2080	359.2080	359.21	362.634	359.2080	359.2080
Function Evaluations	30000	563490	22050	513	252	4515	1710

TABLE 6.21   Statistical Results for the Reinforced Concrete Beam Problem Using CI-SPF, CI-SAPF, CI-SAPF-CBO and CBO.

Results	CI-SPF	CI-SAPF	CI-SAPF-CBO	CBO
Best	359.2080	359.2080	359.2080	362.6340
Mean	359.2083	359.2425	359.3150	379.0192
Worst	359.2094	359.4712	359.9530	386.4000
Std. Dev.	4.61e-4	0.0825	0.1853	9.3515
Avg. CPU Time *sec.*	1.47	0.73	0.47	0.03
Avg. Fun. Eval.	18302	3047	2255	182

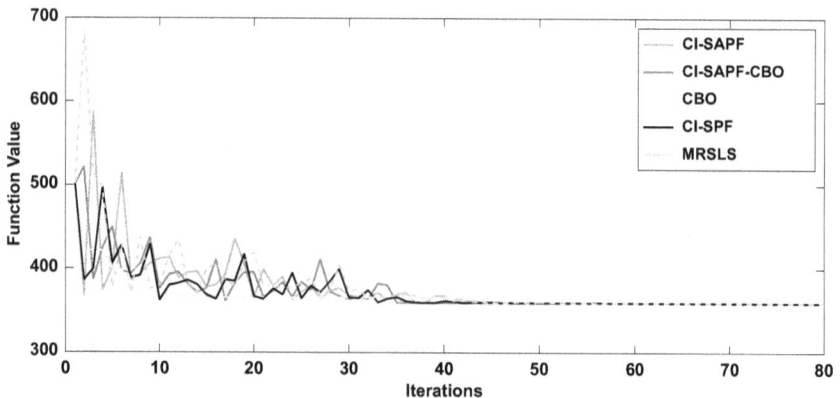

FIGURE 6.12   Comparison of the Convergence Curves of CI-SAPF, CI-SAPF-CBO and CBO for Solving the Reinforced Concrete Beam Design Problem.

TABLE 6.22 Comparison of Results for Solving the Welded Beam Design Case 1 Problem.

Design variables	GA (Deb, 1991)	GA (Coello, 2000)	GA (Coello, and Montes, 2002)	PSO (He and Wang, 2007)	CBO (Kaveh and Mahdavi, 2014)	CI-SPF	MRSLS	CI-SAPF	CI-SAPF-CBO
h	0.248900	0.208800	0.205986	0.202369	0.205722	0.1974	0.1408	0.1773	0.1716
l	6.173000	3.420500	3.471328	3.544214	3.47041	3.0884	4.6724	3.5007	3.6357
t	8.178900	8.997500	9.020224	9.04821	9.037276	9.9988	9.8254	9.9939	9.9998
b	0.253300	0.210000	0.20648	0.205723	0.205735	0.168	0.1740	0.1682	0.1680
$F(x)$	2.433116	1.748310	1.728226	1.728024	1.724663	1.5587	1.6384	1.5510	1.5493
Function Evaluations	NA	NA	NA	NA	NA	5300	5970	11786	9552

TABLE 6.23 Optimal Solutions for the Welding Beam Design Case 2 Problem.

Variables/functions	GA (Deb and Goyal, 1996)	PSO (Eberhart and Kennedy, 1995)	PSO (Datta and Figueira, 2011)	CI-SPF	MRSLS	CBO	CI-SAPF	CI-SAPF-CBO
x_1	1	1	1	1	0	1	1	1
x_2	0	0	0	0	0	1	0	0
h	0.1875	0.125000	0.187500	0.125	0.3125	0.125	0.125	0.125
t	8.2500	8.312500	8.250000	8.25	8.25	15.625	8.25	8.25
b	0.2500	0.250000	0.250000	0.25	0.25	0.75	0.25	0.25
l	1.6849	4.115994	1.782103	1.0003	3.2906	3.7514	1.0000	1.0000
$-g_1$	380.1660 (380.165289)	823.9018	380.1652	380.1653	380.1653	5.2475E+03	380.1653	3.80E+02
$-g_2$	402.0473 (402.047213)	435.7078	402.047	402.0472	402.0472	1.1229E+05	402.0472	4.02E+02
$-g_3$	0.2346 (0.234362)	0.2347	0.2343	0.2343	0.2343	0.2484	0.2343	2.34E-01
$-g_4$	0.1621 (-394.196461)	363.7008	0.0042	6326.1	1.3643	930.1856	6326.1	6.33E+03
$F(x)$	1.9422 (infeasible)	2.0253	1.9553	1.6477	2.0703	4.9773	1.6477	1.6477
Function Evaluations	NA	NA	NA	6600	4592	1152	3000	7092

TABLE 6.24 Statistical Results for the Welded Beam Problem Case 1 and Case 2 Using CI-SAPF, CI-SAPF-CBO and CBO.

Results	Case1			Case2			
	CI-SPF	CI-SAPF	CI-SAPF-CBO	CI-SPF	CI-SAPF	CI-SAPF-CBO	CBO
Best	1.5587	1.5510	1.5492	1.6477	1.6477	1.6477	4.9773
Mean	1.5713	1.5603	1.5527	1.9349	1.8810	1.9332	7.3561
Worst	1.6115	1.5742	1.5591	2.0825	2.0828	2.1552	10.49981
Std. Dev.	0.0154	0.0077	0.0035	0.1791	0.2013	0.2138	2.13821
Avg. CPU Time *sec.*	5.50	1.57	2.24	2.88	2.08	2.15	0.27
Avg. Fun. Eval.	13163	11786	8721	7214	8740	6996	684

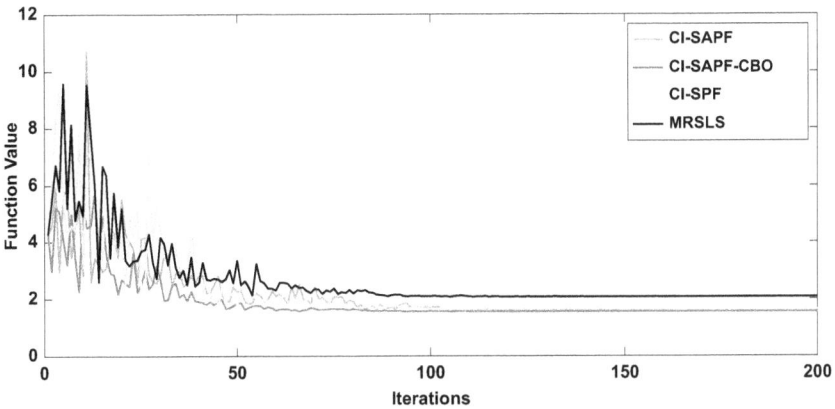

FIGURE 6.13 Comparison of the Convergence Curves of CI-SPF, MRSLS, CI-SAPF, CI-SAPF-CBO and CBO for Solving the Welded Beam Case 1 Design Problem.

from 30 trials using the CI-SPF, CI-SAPF and CI-SAPF-CBO algorithms are presented in Table 6.24. The convergence plots for Case 1 and Case 2 are presented in Figures 6.13 and 6.14, respectively.

6.2.5 Multiple Disc Clutch Brake

The multiple disc clutch brake design problem was previously solved using NSGA (Deb and Srinivasan, 2005), PSO, SBC, BBO, DE and AIA (Rao and Savsani, 2012). These algorithms have been used to successfully validate the more recent CI-SPF, CI-SAPF and CI-SAPF-CBO algorithms by

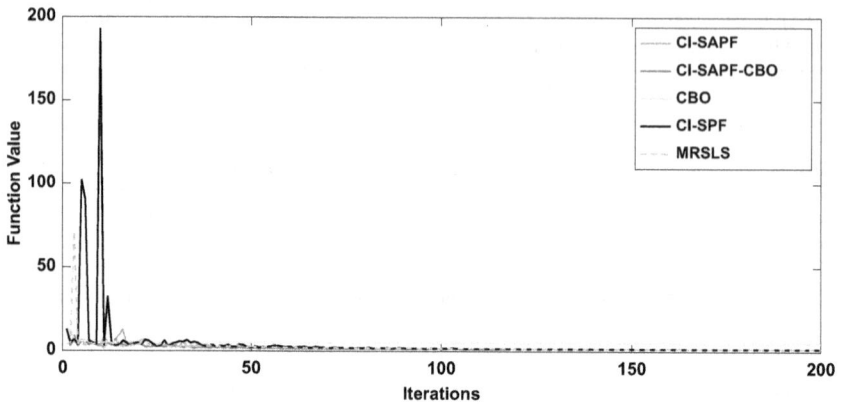

FIGURE 6.14 Comparison of the Convergence Curves of CI-SPF, MRSLS, CI-SAPF, CI-SAPF-CBO and CBO for Solving the Welded Beam Case 2 Design Problem.

comparing their solutions; superior results were obtained with much less computational cost (see Table 6.25). The best, mean and worst function values obtained for CI-SAPF and CI-SAPF-CBO algorithm are similar. The statistical results from 30 trials are presented in Table 6.26, and the convergence curves are presented in Figure 6.15. The solution obtained using CBO was marginally worse.

6.2.6 Helical Compression Spring Design

A mixed variable tension-compression helical spring problem was previously solved in Yun (2005), Gandomi et al. (2011), Sandgren (1990) and Kulkarni et al. (2016a). The solutions obtained using the CI-SPF, CI-SAPF and CI-SAPF-CBO algorithms are successfully validated and compared with other contemporary algorithms, as presented in Table 6.27. It is observed that the results obtained from CI-SAPF were similar to those from FA and PC; however, computationally CI-SAPF performed better with less function evaluations. AHGA yielded the best solution for solving the spring design problem. This was possibly due to its improved iterative hill climbing method associated with a local search. The best, mean and worst function values with standard deviation, average CPU time and average function evaluations are presented in Table 6.28. The convergence curves for CI-SAPF, CI-SAPF-CBO and CBO are presented in Figure 6.16. The other computational details and associated parameters are presented in Table 6.35.

TABLE 6.25 Comparison of Results for Solving the Multiple Disc Clutch Brake Design Problem.

Design Variables	NSGA (Deb and Srinivasan, 2005)	(Rao et al., 2012)									CI-SAPF-
		PSO	ABC	BBO	DE	AIA	CI-SPF	MRSLS	CBO	CI-SAPF	CBO
r_i	70	NA	NA	NA	NA	NA	70	70	66	70	70
r_o	90	NA	NA	NA	NA	NA	90	90	90	90	90
t	1.5	NA	NA	NA	NA	NA	1	1	1	1	1
F	1000	NA	NA	NA	NA	NA	760	760	640	760	760
Z	3	NA	NA	NA	NA	NA	2	2	2	2	2
Mass (kg)	0.4074	0.313657	0.313657	0.313657	0.313657	0.321498	0.2352	0.2352	0.2752	0.2352	0.2352
Function Evaluations	NA	NA	NA	NA	NA	NA	875	450	210	825	450

TABLE 6.26   Statistical Results for the Multiple Disc Clutch Brake Problem Using CI-SPF, CI-SAPF, CI-SAPF-CBO and CBO.

Results	CI-SPF	CI-SAPF	CI-SAPF-CBO	CBO
Best	0.2352	0.2352	0.2352	0.2752
Mean	0.2966	0.2352	0.2352	0.4401
Worst	0.2455	0.2352	0.2352	0.8257
Std. Dev.	0.0036	8.62E-17	9.25E-18	0.2258
Avg. CPU Time *sec.*	0.46	0.23	0.18	0.08
Avg. Fun. Eval.	832	671	800	450

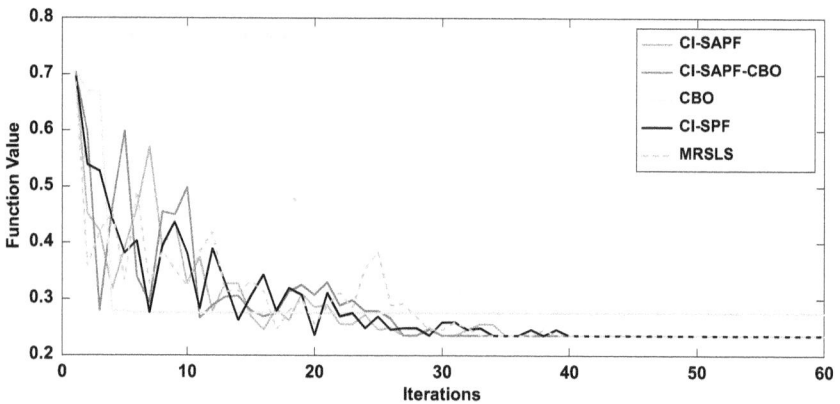

FIGURE 6.15   Comparison of the Convergence Curves of CI-SPF, MRSLS, CI-SAPF, CI-SAPF-CBO and CBO for Solving the Multi-Disc Clutch Design Problem.

6.2.7 I-section Beam Design Problem for Minimisation of Vertical Deflection

In this problem, the CI-SPF, CI-SAPF, CI-SAPF-CBO and CBO algorithms reported better function values than CS (Gandomi et al., 2013) and SOS (Cheng and Prayogo, 2014). A comparison of the results is presented in Table 6.29. The statistical results including best, mean and worst function value with standard deviation, average CPU time and average function evaluations are presented in Table 6.30. For this problem, the standard deviations obtained by CS and SOS were $1.3E-4$ and $4.0E-5$, respectively; however, the CI-SAPF algorithm is more robust with a standard deviation of $4.64E-7$ and less computation cost (see Table 6.27). The convergence curves are shown in Figure 6.17. The other computational details and associated parameters are presented in Table 6.35.

Validation ■ 105

TABLE 6.27 Comparison of Results for Solving the Helical Spring Design Problem.

Design Variables and Constraints	Nonlinear B&B (Sandgren, 1990)	AHGA (Yun, 2005)	FA (Gandomi et, al., 2011)	PC (Kulakrni et al., 2016a)	CI-SPF	MRSLS	CBO	CI-SAPF	CI-SAPF-CBO
d	0.2830	0.2830	0.2830	0.2830	0.283	0.283	2	0.283	0.2830
D	1.180701	1.1096	1.223049	1.2231	1.2233	1.1808	0.3310	1.2230	1.2231
N	10	9	9	9	9	10	4	9	9
g_1	-5430.9	-1.2769e+04	-1008.02	-1.0023	-978.8633	-5413.2530	-1.3378E+04	-1.0066	-1.0023
g_2	-8.8187	-9.1745	-8.946	-0.0089	-8.9444	-8.6484	-10.0602	-0.0089	-0.0089
g_3	-0.08298	-0.0630	-0.083	-0.0001	-0.0830	-0.0830	-0.1312	-0.0001	-0.0001
g_4	-1.8193	-1.890	-1.777	-0.0018	-1.4937	-1.5361	-0.6690	-0.0015	-0.0018
g_5	-1.1723	-1.219	-1.322	-0.0055	-1.3227	-1.1725	-3.0429	-0.0013	-0.0055
g_6	-5.4643	-5.464	-5.464	-0.0055	-5.4639	-5.4642	-5.4436	-0.0055	-0.0055
g_7	0	0	0	0	-0.5361	-0.5357	-0.5544	-0.0005	0
g_8	0	-0.0014	0.0000	0	-8.7709e-04	-6.2911	-0.0482	-0.0000	0
Spring Volume $F(x)\,in^3$	2.7995	2.0283	2.6586	2.6586	2.6592	2.8002	3.2439	2.6586	2.6586
Function Evaluations	NA	NA	NA	498567	3480	2044	108	840	1000

TABLE 6.28   Statistical Results for the Tension-Compression Helical Spring Problem Using CI-SPF, CI-SAPF, CI-SAPF-CBO and CBO.

Results	CI-SPF	CI-SAPF	CI-SAPF-CBO	CBO
Best	2.6892	2.6586	2.6586	3.2440
Mean	2.7425	2.6601	2.6594	4.1223
Worst	2.9100	2.6722	2.6630	5.5331
Std. Dev.	0.0923	0.0032	0.0014	0.8765
Avg. CPU Time *sec.*	2.19	2.05	2.02	0.04
Avg. Fun. Eval.	3241	3030	2991	65

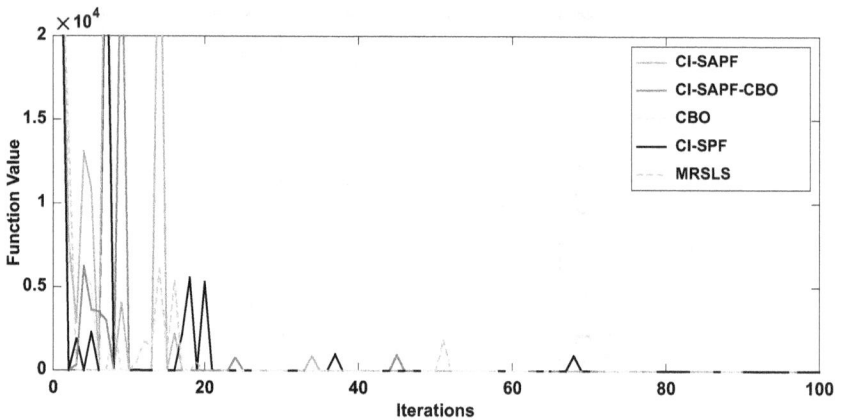

FIGURE 6.16   Comparison of the Convergence Curves of CI-SPF, MRSLS, CI-SAPF, CI-SAPF-CBO and CBO for Solving the Helical Spring Design Problem.

6.2.8 Cantilever Beam

The continuous variable cantilever beam design problem was designed for the minimisation of weight. The CI-SPF, CI-SAPF and CI-SAPF-CBO approaches are successfully validated by solving the problem, and a comparison with other techniques is presented in Table 6.31. The function values obtained using the CI-SPF, CI-SAPF and CI-SAPF-CBO algorithms for solving cantilever beam problem are the same as compared to those from CS (Gandomi et al., 2013) and SOS (Cheng and Prayogo, 2014) and equally robust. The statistical results are presented in Table 6.32. The other computational details and associated parameters are presented in Table 6.35. The convergence curves are shown in Figure 6.18.

TABLE 6.29   Comparison of Results for Solving the I-Section Beam Design Problem for Minimisation of Vertical Deflection.

Variables	CS (Gandomi et al., 2013)	SOS (Cheng and Prayogo, 2014)	CI-SAF	MRSLS	CBO	CI-SAPF	CI-SAPF-CBO
h	80	80	80	80	80	80	50
B	50	50	50	50	50	50	50
t_w	0.90	0.90	1.723	1.7432	1	1.7646	1.7647
t_f	2.3216715	2.32179	4.9962	4.9971	4	4.9999	4.9999
Deflection $F\,min$	0.0130747	0.0130741	0.6628e-2	0.6645e-2	0.8206e-2	0.6626e-2	0.6639e-2
Function Evaluations	5000	5000	2640	686	432	3900	2400

TABLE 6.30   Statistical Results for the I-Section Beam Problem Using CI-SPF, CI-SAPF, CI-SAPF-CBO and CBO.

Results	CI-SPF	CI-SAPF	CI-SAPF-CBO	CBO
Best	0.6628e-2	0.66261e-2	0.6639e-2	0.00820
Mean	0.6662e-2	0.66264e-2	0.6643e-2	0.03843
Worst	0.6749e-2	0.66280e-2	0.6662e-2	0.12922
Std. Dev.	4.54e-5	4.65E-07	9.65E-06	0.0511
Avg. CPU Time $sec.$	1.35	1.27	0.47	0.02
Avg. Fun. Eval.	2453	7830	4553	123

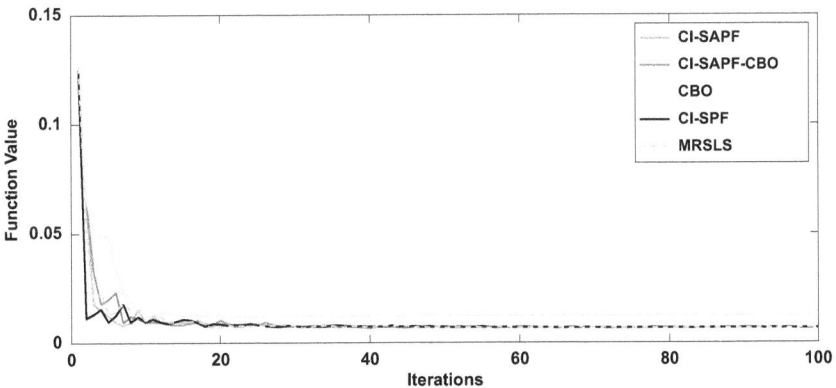

FIGURE 6.17   Comparison of the Convergence Curves of CI-SPF, MRSLS, CI-SAPF, CI-SAPF-CBO and CBO for Solving the I-Section Beam Problem.

TABLE 6.31   Comparison of Results for Solving the Cantilever Beam Design.

Variables	CS (Gandomi et al., 2013)	SOS (Cheng and Prayogo, 2014)	CI-SPF	MRSLS	CBO	CI-SAPF	CI-SAPF-CBO
x_1	6.0089	6.01878	5.974658	5.9356	12.4548	6.0064	6.0090
x_2	5.3049	5.30344	5.326105	5.2700	5.4801	5.3134	5.2944
x_3	4.5023	4.49587	4.562642	4.5587	8.3861	4.4983	4.5074
x_4	3.5077	3.49896	3.472448	3.5333	19.8751	3.4952	3.5097
x_5	2.1504	2.15564	2.165031	2.1932	5.0897	2.1602	2.1531
Weight F min	1.33999	1.33996	1.3400	1.3410	3.2002	1.3399	1.3399
Function Evaluations	NA	15000	3650	680	2190	13750	3025

TABLE 6.32   Statistical Results for the Cantilever Beam Problem Using CI-SPF, CI-SAPF, CI-SAPF-CBO and CBO.

Results	CI-SPF	CI-SAPF	CI-SAPF-CBO	CBO
Best	1.34001	1.3399	1.3399	3.2002
Mean	1.34003	1.3399	1.3400	6.2892
Worst	1.34009	1.3400	1.3402	7.58752
Std. Dev.	3.40e-5	1.74E-07	6.51E-05	1.7810
Avg. CPU Time *sec.*	1.3531	0.9143	1.8353	0.0692
Avg. Fun. Eval.	11371	9581	11761	2320

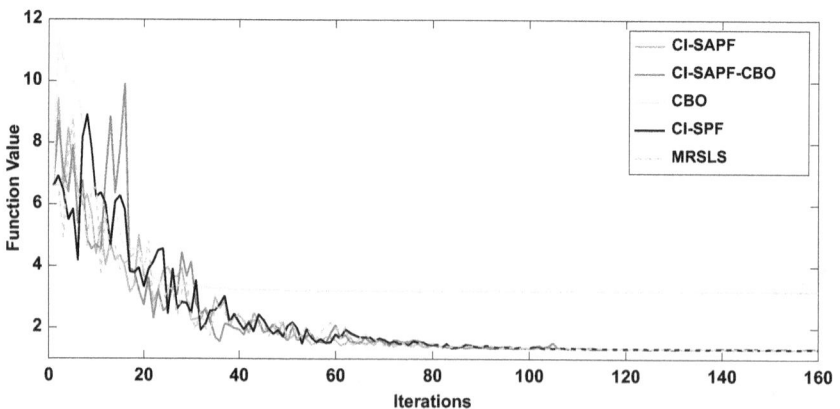

FIGURE 6.18   Comparison of the Convergence Curves of CI-SPF, MRSLS, CI-SAPF, CI-SAPF-CBO and CBO for Solving the Cantilever Beam Problem.

6.2.9 Compound Gear Train

For the compound gear train design problem, the CI-SPF, CI-SAPF, CI-SAPF-CBO and CBO algorithms are successfully validated and compared with the methods such as the nonlinear B&B (Sandgren, 1990), Lagrange multiplier (Kannan and Kramer, 1994) and PSO (Datta and Figueira, 2011), as presented in Table 6.33. The CI-SPF, CI-SAPF and CI-SAPF-CBO algorithms obtained similar results as the PSO (former best solution). The best, mean and worst solutions obtained using CI-SAPF, CI-SAPF-CBO and CBO with standard deviation, average CPU time and average function evaluations are illustrated in Table 6.34. The other computational details and associated parameters are presented in Table 6.35. The convergence curves are shown in Figure 6.19

6.3 LINEAR AND NONLINEAR BENCHMARK TEST PROBLEMS

The CI-SAPF, CI-SAPF-CBO and CBO algorithms are also applied for solving several maximisation and minimisation test problems (Srivastava and Fahim, 2001), including the dynamic variable, transportation, multi-stage, Rosen–Suzuki convex test, knapsack and two cases of integer linear problems. Moreover, two cases of non-convex integer problem and global nonlinear mixed discrete programming problem (Tsai et al., 2002), the 3-bar truss test problem (Shin et al., 1990) and six monotone functions (Lawler and Bell, 1966) are also solved. All these problems consist of discrete variables and linear-nonlinear type constraints. The discrete variables are handled using a round off integer sampling technique (Kale and Kulkarni, 2018) and constraints are handled using a proposed SAPF approach. CBO could not achieve the feasible solution for the transportation problem and four monotone functions. The convergence curves for the test problems are illustrated in Figures 6.20 to 6.34. The proposed methods exhibited comparative and similar results as that of contemporary approaches. For all the solved problems the comparison of results is presented in Table 6.36. The problem statements of these problems are illustrated in Appendix.

6.4 RESULTS, ANALYSIS AND DISCUSSION

Discrete variable truss structural optimisation problems, mixed variable mechanical engineering design problems and linear and nonlinear optimisation test problems are successfully solved using CI-SPF, CI-SAPF and CI-SAPF-CBO algorithms, and the results are compared with other

TABLE 6.33 Comparison of Results for Solving the Compound Gear Design Problem.

Variables	Nonlinear B&B (Sandgren, 1990)	Lagrange Multiplier (Kannan and Kramer, 1994)	PSO (Datta and Figueira, 2011)	CI-SPF	MRSLS	CBO	CI-SAPF	CI-SAPF-CBO
z_a	18	13	16	16	17	18	16	16
z_b	22	15	19	19	27	20	19	19
z_c	45	33	43	43	60	58	43	43
z_d	60	41	49	49	53	43	49	49
$f(x)$	5.7e-06	2.1246e-08	2.7e-12	2.70e-12	1.81e-05	4.5033E-9	2.70e-12	2.70e-12
Gear ratio	0.146666	0.1441	0.144281	0.144284	0.144339	0.144346	0.144281	0.144281
Error %	1.65 %	0.11 %	0.0011 %	0.0011 %	1.67 %	0.0462 %	0.0011 %	0.0011 %
Function Evaluations	NA	NA	NA	4200	242	420	1260	360

TABLE 6.34   Statistical Results for the Compound Gear Train Problem Using CI-SPF, CI-SAPF, CI-SAPF-CBO and CBO.

Results	CI-SPF	CI-SAPF	CI-SAPF-CBO	CBO
Best	2.7E-12	2.7E-12	2.7E-12	4.50E-09
Mean	3.40E-06	3.97E-11	1.73E-11	1.46E-05
Worst	1.65E-05	6.60E-10	2.31E-11	7.06E-05
Std. Dev.	6.20952E-06	5.26E-12	9.55E-12	3.13475E-05
Avg. CPU Time *sec.*	1.5891	3.1249	2.1784	0.037685083
Avg. Fun. Eval.	4329	8513	6358	102

contemporary techniques. These problems are also solved using the CBO algorithm in order to compare the performance of individual algorithms. The problems considered here are 10 discrete variable truss structure problems, 11 mixed variable design engineering problems and 17 linear and nonlinear test problems with integer variables. For every individual problem, the CI-SPF, MRSLS, CI-SAPF, CI-SAPF-CBO and CBO algorithm is run 30 times with different initialisations for every learning attempt. The statistical results of all the problems considered in the work are presented in Table 6.35 and Table 6.36. Table 6.35 presents a statistical comparison of the best, mean and worst function values for CI-SPF, CI-SAPF and CI-SAPF-CBO including standard deviation of function values, average number of function evaluations, average computational time, closeness to the reported solution and the set of parameters required to run the CI-SPF and CI-SAPF algorithms. Table 6.36 presents the comparison of solutions obtained from the CI-SPF, MRSLS, CI-SAPF, CI-SAPF-CBO and CBO algorithms along with formerly used techniques for solving linear and non-linear optimisation test problems. It is observed that solutions obtained using CI-SPF are computationally (CPU time and function evaluations) efficient as well as robust compared to other methods. Moreover, MRSLS has also been successfully applied to these problems. It is important to mention that MRSLS obtained better solutions than CI-SPF, especially for the 25-bar (Case 2) (see Table 6.5) and 72-bar (Case 2) (see Table 6.14) truss structure problems. However, it is unable to obtain better solutions for the remaining problems studied in this book. This is because for every MRSLS run, a solution is randomly initialised that could be infeasible in which case, MRSLS might be able to find a better solution.

In the CI-SAPF-CBO algorithm, CBO enhanced the exploration of search space, enabling the CI algorithm to reach towards better solutions

TABLE 6.35   Comparison of Statistical Results for the Truss Structure and Engineering Design Problems Using CI-SPF, CI-SAPF and CI-SAPF-CBO Algorithms.

Test Examples	So far best reported solution	Proposed Techniques	CI-SPF solution best mean worst	Standard Deviation of System sol.	Average number of function evaluations
6-Bar Problem	4962.1 (GA)	CI-SPF	4962.0966 4962.0966 4962.0966	0	3831
		CI-SAPF	4962.0966 4962.0966 4962.0966	0	2735
		CI-SAPF-CBO	4962.0966 4962.0966 4962.0966	0	2449
Case 1 of 10-Bar Problem	5490.74 (ADS)	CI-SPF	5490.7387 5501.1295 5508.3824	12.3321	15813
		CI-SAPF	5490.6020 5505.7540 5534.9590	16.6546	22654
		CI-SAPF-CBO	5490.6021 5505.7142 5532.8502	14.0915	16845
Case 2 of 10-Bar Problem	4686.7729 (PC)	CI-SPF	5062.1097 5064.7229 5067.3314	2.5353	23383
		CI-SAPF	5061.7600 5062.2160 5067.3310	1.5402	22874
		CI-SAPF-CBO	5061.7597 5062.5006 5067.3314	1.8895	24325
Case 1 of 25-Bar Problem	477.1668 (PC)	CI-SPF	516.2025 545.3048 557.6885	13.1198	29711
		CI-SAPF	512.8140 530.8760 548.6147	10.1639	22588
		CI-SAPF-CBO	512.8140 531.9475 543.0734	8.8008	21652

Standard Deviation of function evaluations	Average computational time (*sec.*)	Standard Deviation of computational time	Closeness to the best reported solution %	Set of parameters C, R, t, θ
223.1432	4.58	1.2745	0.00	$C = 7$, $R = 0.97$, $t = 15$, $\theta = 6000$
237.6711	3.73	1.2712	00.00	$C = 5$, $R = 0.965$
1139.3211	3.32	1.9992	0.00	$C = 6$
7431.2147	12.95	7.9654	00.0	$C = 7$, $R = 0.97$, $t = 15$, $\theta = 6000$
8346.8900	17.87	8.5247	0.0024	$C = 5$, $R = 0.97$
7763.6032	10.81	4.8338	0.0024	$C = 6$
13712.6441	32.56	18.4355	8.00*	$C = 7$, $R = 0.97$, $t = 15$, $\theta = 6000$
17365.79	31.72	23.3469	8.00*	$C = 5$, $R = 0.975$
15622.9321	34.53	21.6393	8.00*	$C = 6$
12535.5056	17.66	7.2302	8.18*	$C = 7$, $R = 0.97$, $t = 15$, $\theta = 11000$
9028.3586	12.22	4.8828	7.4705*	$C = 5$, $R = 0.97$
8547.8230	11.71	4.6229	7.47*	$C = 6$

(*continued*)

TABLE 6.35 Comparison of Statistical Results for the Truss Structure and Engineering Design Problems using CI-SPF, CI-SAPF and CI-SAPF-CBO Algorithms. (Continued)

Test Examples	So far best reported solution	Proposed Techniques	CI-SPF solution best mean worst	Standard Deviation of System sol.	Average number of function evaluations
Case 2 of 25-Bar Problem	464.1471 (PC)	CI-SPF	500.0448 521.1982 546.1536	12.7871	23002
		CI-SAPF	473.4724 486.7399 498.0590	6.8008	26796
		CI-SAPF-CBO	473.4725 485.9673 487.9375	2.8152	28513
38-Bar Problem	5889.9 (ISCSO)	CI-SPF	5900.3465 5916.9868 5938.0018	10.52438	89801
		CI-SAPF	5891.0472 5895.3762 5898.4401	2.1872	69920
		CI-SAPF-CBO	5889.9511 5903.2514 5927.0836	10.9272	46900
45-Bar Problem	14341.21 (ISCSO)	CI-SPF	14354.2652 14390.4386 14410.2133	16.7597	131378
		CI-SAPF	14322.2890 14455.6812 14667.1000	91.7773	114525
		CI-SAPF-CBO	14322.2890 14413.8947 14480.4395	40.1861	103237
52-Bar Problem	1899.35 (CBO)	CI-SPF	1894.4817 1929.7643 1990.3652	27.0146	124581
		CI-SAPF	1894.4817 1913.9740 1934.94032	11.9890	101751
		CI-SAPF-CBO	1891.4391 1909.3269 1920.1055	7.1767	95645
Case 1 of 72-Bar Problem	372.4095 (PC)	CI-SPF	373.5427 385.6413 391.6090	5.1138	94012
		CI-SAPF	372.4095 381.5319 384.6292	3.2536	85500

Standard Deviation of function evaluations	Average computational time (*sec.*)	Standard Deviation of computational time	Closeness to the best reported solution %	Set of parameters C, R, t, θ
16181.1000	14.23	9.1015	7.73*	$C=7$, $R=0.95$, $t=15$, $\theta=7000$
10818.7571	15.95	6.4404	2.009*	$C=5$, $R=0.95$
15195.7166	22.18	11.8218	2.0091*	$C=6$
41874.8601	42.30	19.8588	0.177*	$C=7$, $R=0.9678$, $t=15$, $\theta=6800$
35254.8125	40.60	20.4629	0.0019*	$C=5$, $R=0.978$
10066.8617	33.30	7.5267	0	$C=6$
59986.1900	80.45	35.1687	0.0911*	$C=7$, $R=0.974$, $t=15$, $\theta=12000$
45601.0400	63.17	25.1541	0.1319	$C=5$, $R=0.98$
56672.6423	56.94	31.2613	0.1319	$C=6$
65747.6643	71.33	42.2731	0.25	$C=7$, $R=0.98$, $t=15$, $\theta=7500$
46869.9796	52.73	23.8245	0.2563	$C=5$, $R=0.973$
37590.9291	61.16	24.0402	0.4165	$C=6$
50620.51	64.36	33.7853	0.30*	$C=7$, $R=0.966$, $t=15$, $\theta=9000$
36751.2800	78.61	31.3196	0.00	$C=5$, $R=0.977$

(*continued*)

TABLE 6.35 Comparison of Statistical Results for the Truss Structure and Engineering Design Problems using CI-SPF, CI-SAPF and CI-SAPF-CBO Algorithms. (Continued)

Test Examples	So far best reported solution	Proposed Techniques	CI-SPF solution best mean worst	Standard Deviation of System sol.	Average number of function evaluations
		CI-SAPF-CBO	372.4095 398.0869 418.9014	11.8925	85872
Case 2 of 72-Bar Problem	379.9079 (PC)	CI-SPF	381.3418 385.8209 395.1080	3.1475	181496
		CI-SAPF	380.1807 383.2420 385.9565	1.6929	124080
		CI-SAPF-CBO	379.0668 387.8078 397.3978	5.4302	91872
Pressure Vessel Design Problem	6059.7143 (OIO)	CI-SPF	6059.7152 6069.6323 6091.6584	12.6409	13023
		CI-SAPF	6051.4819 6065.2506 6103.2707	11.91071208	11974
		CI-SAPF-CBO	6059.7151 6065.3150 6090.5326	9.2158	12114
Stepped Cantilever Beam Design Problem	61473.93 (PC)	CI-SPF	63893.4914 63905.8800 63946.7700	606.06	20174
		CI-SAPF	63893.4544 64248.9590 64333.4510	175.2922	21552
		CI-SAPF-CBO	63893.4682 64229.2335 64549.4303	207.1399	30437
Speed Reducer Problem	2828.5863 (PC)	CI-SPF	2817.5581 2824.9781 2828.4442	2.7850	10851
		CI-SAPF	2817.0910 2820.3190 2825.8710	2.0660	14420
		CI-SAPF-CBO	2816.3410 2817.6529 2819.5744	1.0276	7597

Standard Deviation of function evaluations	Average computational time (*sec.*)	Standard Deviation of computational time	Closeness to the best reported solution %	Set of parameters C, R, t, θ
69269.4621	53.88	42.4913	0.00	$C = 6$
71257.89	131.18	54.3689	0.377*	$C = 7, \ R = 0.969, \ t = 15, \ \theta = 9000$
56364.8812	118.52	57.4261	0.0718*	$C = 5, \ R = 0.969$
36459.9623	80.39	31.9038	0.2213	$C = 6$
6575.1859	4.93	3.2209	1.485e10-5*	$C = 7, \ R = 0.955, \ t = 15, \ \theta = 10000$
7517.6672	4.43	2.6297	0.1358	$C = 5, \ R = 0.965$
5286.2451	4.29	0.8249	1.32e-5*	$C = 6$
6320.326	4.84	1.8274	3.936*	$C = 7, \ R = 0.92, \ t = 15, \ \theta = 5500$
12383.9200	5.36	3.292	2.6450*	$C = 5, \ R = 0.952$
9753.7751	7.45	2.4845	3.9358*	$C = 6$
4692.52783	2.69	1.4213	0.389	$C = 7, \ R = 0.935, \ t = 15, \ \theta = 7000$
6404.4916	2.25	0.9134	0.4064	$C = 5, \ R = 0.946$
6387.8979	2.01	1.6956	0.0432	$C = 6$

(continued)

TABLE 6.35 Comparison of Statistical Results for the Truss Structure and Engineering Design Problems using CI-SPF, CI-SAPF and CI-SAPF-CBO Algorithms. (Continued)

Test Examples	So far best reported solution	Proposed Techniques	CI-SPF solution best mean worst	Standard Deviation of System sol.	Average number of function evaluations
Reinforced Concrete Beam Problem	359.2080 (FA, PC)	CI-SPF	359.2080 359.2084 359.2095	0.00046	18301
		CI-SAPF	359.2080 359.2425 359.4712	0.082515	3047
		CI-SAPF-CBO	359.2080 359.3150 359.9530	0.1853	2255
Case 1 of Welded Beam Problems	1.7247 (CBO)	CI-SPF	1.5587 1.5713 1.6115	0.0154	13163
		CI-SAPF	1.5510 1.5603 1.5742	0.007716137	11786
		CI-SAPF-CBO	1.5493 1.5527 1.5591	0.0035	8721
Case 2 of Welded Beam Problem	1.9553 (PSO)	CI-SPF	1.6477 1.9349 2.0828	0.1792	7214
		CI-SAPF	1.6477 1.8810 2.0828	0.2013	8740
		CI-SAPF-CBO	1.6477 1.9332 2.1552	0.2 138	6996
Multiple Disc Clutch Problem	0.3136 (DE)	CI-SPF	0.2352 0.2366 0.2455	0.0036	832
		CI-SAPF	0.2352 0.2352 0.2352	8.62e-17	671
		CI-SAPF-CBO	0.2352 0.2352 0.2352	6.52E-16	800
Tension Compression Helical Spring Problem	2.0283 (AHGA)	CI-SPF	2.6592 2.7425 2.9100	0.0966	3241
		CI-SAPF	2.6586 2.6601 2.6722	0.0032	3030

Standard Deviation of function evaluations	Average computational time (*sec.*)	Standard Deviation of computational time	Closeness to the best reported solution %	Set of parameters C, R, t, θ
9249.3586	1.47	0.8159	00.0	$C = 7$, $R = 0.93$, $t = 15$, $\theta = 5500$
1866.4160	0.73	0.4472	0.00	$C = 5$, $R = 0.93$
1506.5044	0.47	0.3053	0.00	$C = 6$
7293.1968	5.50	3.0485	9.624	$C = 7$, $R = 0.965$, $t = 15$, $\theta = 1000$
5894.6691	4.93	2.4639	10.0713	$C = 5$, $R = 0.965$
6214.0242	2.24	1.5966	10.1698	$C = 6$
4808.6594	2.88	1.9225	15.629	$C = 7$, $R = 0.965$, $t = 15$, $\theta = 1000$
4524.8112	2.08	1.8091	15.7316	$C = 5$, $R = 0.965$
6223.9616	2.15	2.0736	17.8349	$C = 6$
27.6253	0.46	0.0153	25	$C = 7$, $R = 0.952$, $t = 15$, $\theta = 5500$
5.6314	0.23	0.0045	25.00	$C = 5$, $R = 0.952$
83.8525	0.18	0.0558	25.00	$C = 6$
860.0609	2.19	0.5827	31.10*	$C = 7$, $R = 0.96$, $t = 15$, $\theta = 6000$
1423.2952	2.05	0.9644	31.0752*	$C = 5$, $R = 0.96$

(*continued*)

TABLE 6.35   Comparison of Statistical Results for the Truss Structure and Engineering Design Problems Using CI-SPF, CI-SAPF and CI-SAPF-CBO Algorithms. (Continued)

Test Examples	So far best reported solution	Proposed Techniques	CI-SPF solution best mean worst	Standard Deviation of System sol.	Average number of function evaluations
		CI-SAPF-CBO	2.6586 2.6594 2.6630	0.0014	2991
I Section Beam Design Problem	0.0131 (SOS)	CI-SPF	0.6628e-2 0.6662e-2 0.6749e-2	4.5356E-05	2453
		CI-SAPF	0.66261e-2 0.66264e-2 0.66280e-2	4.65e-07	7830
		CI-SAPF-CBO	0.6639e-2 0.6643e-2 0.6662e-2	9.65729E-06	4553
Cantilever Beam Problem	1.3399 (SOS)	CI-SPF	1.34001 1.34003 1.34009	3.4011E-05	11371
		CI-SAPF	1.3399 1.3399 1.3400	1.74e-05	9581
		CI-SAPF-CBO	1.3399 1.3400 1.3402	6.5143E-05	11761
Compound Gear Train Design Problem	2.7E−12 (PSO)	CI-SPF	2.7e−12 3.40e-06 1.65e-05	6.2095E-05	4329
		CI-SAPF	2.7e-12 3.97e-11 6.60e-10	5.2614e-12	8513
		CI-SAPF-CBO	2.7e-12 1.73e-11 2.31e-11	9.553E-12	6358

Note: * Indicates that the solution obtained using CI-SPF, CI-SAPF and CI-SAPF-CBO was worse than other algorithms.

within substantially less computational time and function evaluations. The SAPF approach served to handle the linear and nonlinear constraints, and CI worked as a global optimiser. Most importantly, CI-SAPF-CBO is a generalised algorithm that does not require fine tuning of parameters. This makes it easier to apply the proposed algorithm to a wide range of applications.

Standard Deviation of function evaluations	Average computational time (sec.)	Standard Deviation of computational time	Closeness to the best reported solution %	Set of parameters C, R, t, θ
1423.2952	2.02	0.9644	31.0752*	$C = 6$
350	0.40	0.0573	49.46	$C = 7$, $R = 0.957$, $t = 15$, $\theta = 5500$
5735.1212	1.27	0.9374	49.6183	$C = 5$, $R = 0.957$
2509.8982	0.47	0.2595	49.5100	$C = 6$
5227	1.35	0.6235	0.00	$C = 7$, $R = 0.957$, $t = 15$, $\theta = 5500$
4296.6242	0.91	0.512	0	$C = 5$, $R = 0.957$
5286.2451	1.83	0.8249	0.00	$C = 6$
123.7201	1.58	0.0454	0.00	$C = 7$, $R = 0.9843$, $t = 15$, $\theta = 5500$
6250.8041	3.12	2.2944	0	$C = 5$, $R = 0.9843$
5546.9959	2.17	1.9004	0.00	$C = 6$

The discrete variables truss structure domain problems considered in the present work are solved for minimisation of weight subject to stress (in links) and deflection (in nodes). For 38-bar, 52-bar and 72-bar (Case 1) truss structure problems, CI-SAPF-CBO obtained better results as compared to the CI-SPF, CI-SAPF and CBO algorithms. The CBO algorithm is incorporated with a round off approach to handle the discrete variables and a static penalty function approach to handle the inequality constraints. In the current work, the CI-SAPF and CI-SAPF-CBO algorithms revealed

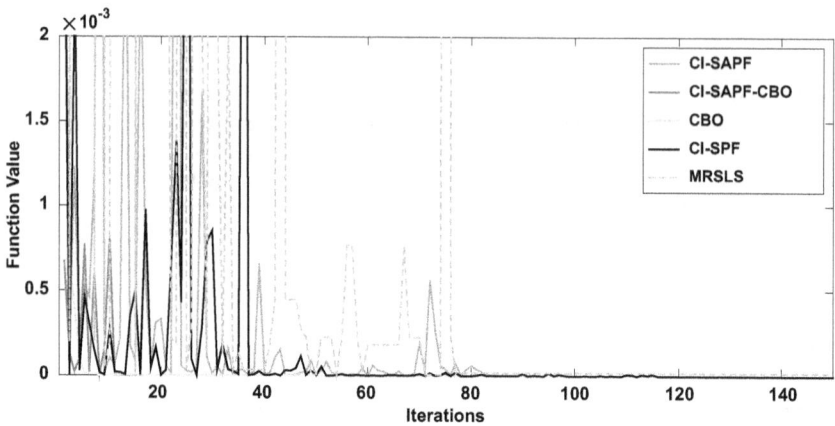

FIGURE 6.19 Comparison of the Convergence Curves of CI-SPF, MRSLS, CI-SAPF, CI-SAPF-CBO and CBO for Solving the Gear Design Problem.

their superiority in dealing with such problems compared to the CI-SPF, CBO (Kaveh and Mahdavi, 2015) and other contemporary algorithms. The results obtained from CI-SAPF and CI-SAPF-CBO are similar, as in both approaches the CI worked as a global optimiser. The proposed SAPF approach handled the constraints and CBO was incorporated to refine the solutions obtained using CI. It is noted that CBO searched the solution unidirectionally to find the best cost function which makes the solution become trapped into the local minima and unable to escape. In CI-SAPF-CBO, a probabilistic roulette wheel approach is associated with CI to assist the algorithm to escape from local minima (see Figures 6.1–6.34).

An adaptive penalty function approach was proposed with the similar motivation to avoid the penalty parameter setting (Nanakorn and Meesomklin, 2001). It was used to solve the 6-bar truss structure problem. In this approach, the penalty parameter was set using the ratio between the best infeasible fitness value and the average feasible fitness value. This process required a separate scaling factor that adjusted the strength of penalty parameter. This could be one of the limitations of adaptive penalty function approach in that it may increase the number of preliminary trials needed to set the scaling factor. In CI-SPF a static penalty function is used to handle constraints that also require preliminary trials to set the penalty parameter. However, in the proposed SAPF approach, the function value is used as a penalty parameter and it updates after every iteration. Ultimately, the function value aimed towards minimisation

TABLE 6.36   Comparison of Results for the Linear and Nonlinear Problems Using the CI-SPF, MRSLS, CI-SAPF, CI-SAPF-CBO and CBO Algorithms.

Sr. No.	Test Functions	Solver	Search Space		Function Value	Optimum variables
			Lower limit	Upper limit		
1	Dynamic Problem (Maximisation)	Srivastava and Fahim (2001)	[0,0,0]	[10,10,10]	16	[0,1,2]
		CI-SPF			16	[0,1,2]
		MRSLS			16	[0,1,2]
		CI-SAPF			16	[0,1,2]
		CI-SAPF-CBO			16	[0,1,2]
		CBO			15	[0,2,1]
2	Transportation Problem	Srivastava and Fahim (2001)	[0,...,0]	[100,...,100]	40.5	[5,15,0,0,0,15,0,5,5]
		CI-SPF			40.8	[2,18,0,3,0,12,0,2,8]
		MRSLS			40.8	[2,18,0,3,0,12,0,2,8]
		CI-SAPF			40.8	[2,18,0,3,0,12,0,2,8]
		CI-SAPF-CBO			40.8	[2,18,0,3,0,12,0,2,8]
		CBO			40.8	[2,18,0,3,0,12,0,2,8]
3	Multistage Problem (Maximisation)	Srivastava and Fahim (2001)	[0,0,0]	[100,100,100]	55.2	[3,1,0]
		CI-SPF			55.2	[3,1,0]
		MRSLS			55.2	[3,1,0]
		CI-SAPF			55.2	[3,1,0]
		CI-SAPF-CBO			55.2	[3,1,0]
		CBO			32.7	[0,1,1]

(continued)

TABLE 6.36 Comparison of Results for the Linear and Nonlinear Problems Using the CI-SPF, MRSLS, CI-SAPF, CI-SAPF-CBO and CBO Algorithms. (Continued)

Sr. No.	Test Functions	Solver	Search Space		Function Value	Optimum variables
			Lower limit	Upper limit		
4	Knapsack Problem (Maximisation)	Srivastava and Fahim (2001)	[0, …,0]	[100, …,100]	19979	[32,2,1,0,0,0,0]
		CI-SPF			20059	[16,18,9,0,2,8,0]
		MRSLS			20059	[16,18,9,0,2,8,0]
		CI-SAPF			20059	[16,18,9,0,2,8,0]
		CI-SAPF-CBO			20240	[27,3,4,4,8,1,5]
		CBO			18428	[16,10,8,0, 19,8,38]
5	Integer Linear Programming (Minimisation)	Srivastava and Fahim (2001)	[0,0]	[100,100]	33	[3,6]
		CI-SPF			33	[3,6]
		MRSLS			33	[3,6]
		CI-SAPF			33	[3,6]
		CI-SAPF-CBO			33	[3,6]
		CBO			33	[3,6]
6	Non-convex Integer problem (formulation 1)	Tsai et al. (2002)	[1,1,1]	[5,5,5]	-75.7579	[1,2,5]
		CI-SPF			-75.7579	[1,2,5]
		MRSLS			-75.7579	[1,2,5]
		CI-SAPF			-75.7579	[1,2,5]
		CI-SAPF-CBO			-75.7579	[1,2,5]
		CBO			-75.7579	[1,2,5]
	(formulation 2)	Tsai et al. (2002)	[0,0,0]	[5,5,5]	-125	[5,4,0]
		CI-SPF			-328.3159	[0,5,5]
		MRSLS			-328.3159	[0,5,5]
		CI-SAPF			-328.3159	[0,5,5]
		CI-SAPF-CBO			-328.3159	[0,5,5]
		CBO			-131.3264	[0,2,5]

No.	Problem	Method			Result	Solution
7	Global nonlinear mixed discrete programming	Tsai et al. (2002)	[3,3]	[6,5]	-246	[5,4]
		CI-SPF			-275	[5,5]
		MRSLS			-275	[5,5]
		CI-SAPF			-275	[5,5]
		CI-SAPF-CBO			-275	[5,5]
		CBO			-275	[5,5]
8	Three-bar truss design problem	Shin et al. (1990)	[0.1, 0.2, 0.3, 0.5, 0.8, 1.0, 1.2]		3.0414	[1.2,0.5,0.1]
		CI-SPF			3.0414	[1.2,0.5,0.1]
		MRSLS			3.0414	[1.2,0.5,0.1]
		CI-SAPF			3.0414	[1.2,0.5,0.1]
		CI-SAPF-CBO			3.0414	[1.2,0.5,0.1]
		CBO			3.0414	[1.2,0.5,0.1]
9	Monotone Functions	Lawler and Bell (1966)	[0,0,0,0,0]	[3,3,3,3,3]	8	[1,1,1,2]
		CI-SPF			16	[1,1,2,1,3]
		MRSLS			16	[1,1,2,1,3]
		CI-SAPF			16	[1,1,2,1,3]
		CI-SAPF-CBO			16	[1,1,2,1,3]
		CBO			16	[1,1,2,1,3]
10		Lawler and Bell (1966)	[0,0,0,0,0,0]	[7,7,7,15,15,7,15]	16	[1,4,1,0,2,1,2]
		CI-SPF			14	[1,4,1,0,2,1,2]
		MRSLS			15	[0,3,3,0,2,1,4]
		CI-SAPF			14	[0,2,4,0,2,1,6]
		CI-SAPF-CBO			14	[0,2,4,0,2,1,6]
		CBO			22	[2,3,1,1,2,1,2]
11		Lawler and Bell (1966)	[0,0,0,0,0,0]	[7,7,7,15,15,7,15]	10	[0,6,0,1,1,1,1]
		CI-SPF			11	[2,4,0,1,1,1,2]
		MRSLS			11	[2,4,0,1,1,1,2]
		CI-SAPF			11	[2,4,0,1,1,1,2]
		CI-SAPF-CBO			11	[2,4,0,1,1,1,2]

(continued)

TABLE 6.36 Comparison of Results for the Linear and Nonlinear Problems Using the CI-SPF, MRSLS, CI-SAPF, CI-SAPF-CBO and CBO Algorithms. (Continued)

Sr. No.	Test Functions	Solver	Search Space		Function Value	Optimum variables
			Lower limit	Upper limit		
12		Lawler and Bell (1966)	[0,0,0,0,0,0,0]	[7,7,7,15,15,7,15]	46	[0,7,0,0,2,1]
		CI-SPF			92	[2,4,0,0,1,2,2]
		MRSLS			92	[2,4,0,0,1,2,2]
		CI-SAPF			92	[2,4,0,0,1,2,2]
		CI-SAPF-CBO			92	[2,4,0,0,1,2,2]
13		Lawler and Bell (1966)	[0,0,0,0,0,0,0]	[7,7,7,15,15,7,15]	25	[1,4,1,1,1,1,2]
		CI-SPF			21	[2,3,1,1,1,1,2]
		MRSLS			21	[2,3,1,1,1,1,2]
		CI-SAPF			21	[2,3,1,1,1,1,2]
		CI-SAPF-CBO			21	[2,3,1,1,1,1,2]
14		Lawler and Bell (1966)	[0,0,0,0,0,0,0]	[7,7,7,15,15,7,15]	1000	[0,7,0,0,2,1]
		CI-SPF			1331	[1,3,2,1,1,1,2]
		MRSLS			1331	[1,3,2,1,1,1,2]
		CI-SAPF			1331	[1,3,2,1,1,1,2]
		CI-SAPF-CBO			1331	[1,3,2,1,1,1,2]

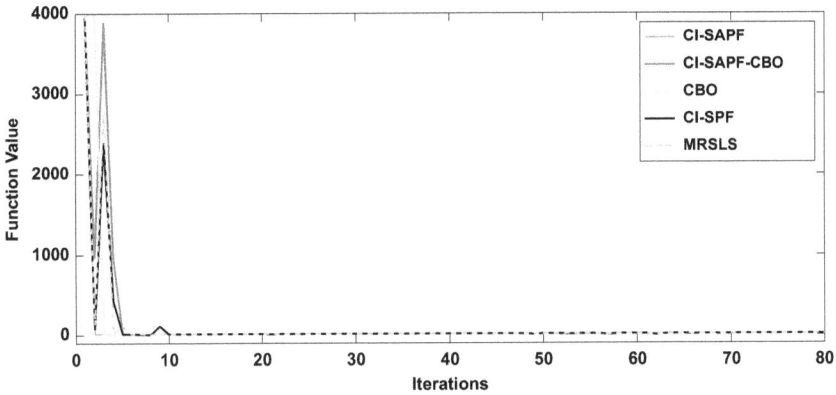

FIGURE 6.20 Comparison of the Convergence Curves of CI-SPF, MRSLS, CI-SAPF, CI-SAPF-CBO and CBO for Solving the Dynamic Programming (Maximisation) Problem.

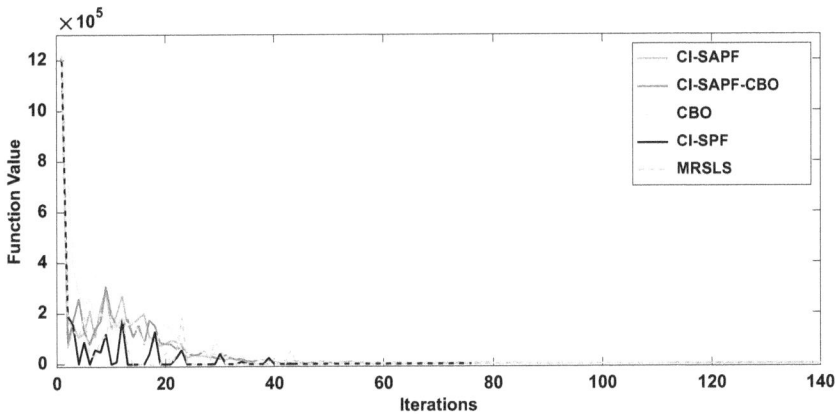

FIGURE 6.21 Comparison of the Convergence Curves of CI-SPF, MRSLS, CI-SAPF, CI-SAPF-CBO and CBO for Solving the Transportation Problem.

which accelerates the convergence rate and obtain competent results with a reduced computational cost. For the 10-bar (Case 2), 25-bar (Case 1 and Case 2), 52-bar and 72-bar 3D spatial discrete truss structure, the CI-SPF, CI-SAPF and CI-SAPF-CBO algorithms performed better than PSO and PSOPC. In PSO and PSOPC (Li et al., 2009), a penalty function was incorporated that then degenerated the function value. To overcome this limitation, a fly-back mechanism was used in HPSO (Li et al., 2009) to handle the constraints which expedited the convergence rate. The 10-bar

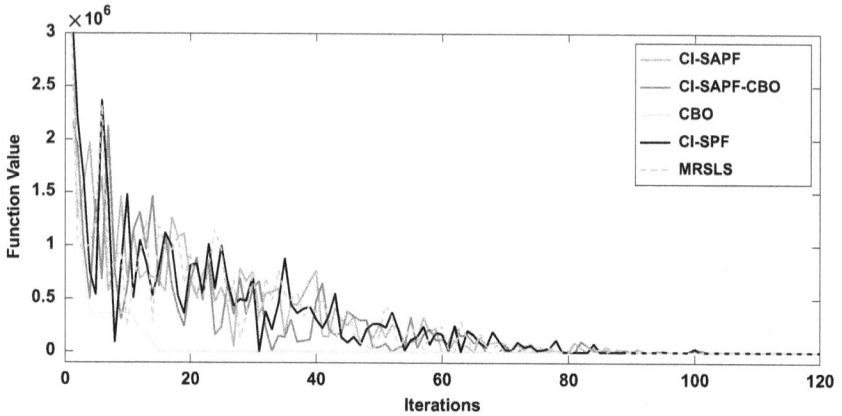

FIGURE 6.22 Comparison of the Convergence Curves of CI-SPF, MRSLS, CI-SAPF, CI-SAPF-CBO and CBO for Solving the Multistage (Maximisation) Problem.

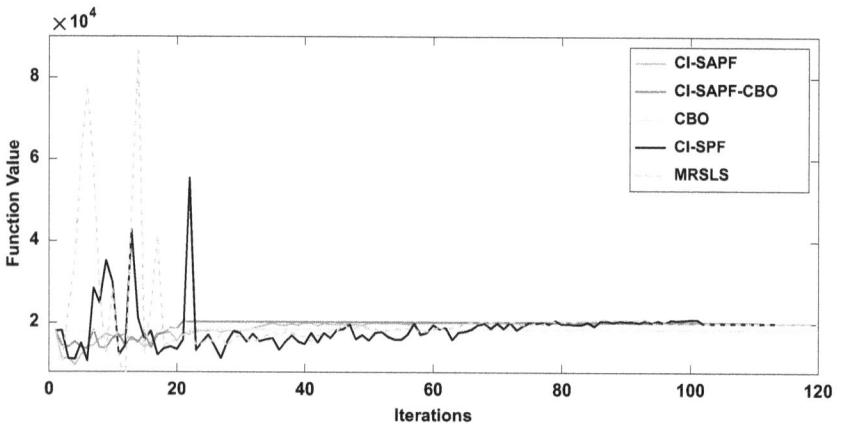

FIGURE 6.23 Comparison of the Convergence Curves of CI-SPF, MRSLS, CI-SAPF, CI-SAPF-CBO and CBO for Solving the Knapsack Problem.

truss structure problem was solved using the steady state GA (Wu and Chow, 1995; Rajeev and Krishnamoorthy, 1992). In (Wu and Chow, 1995), the discrete design variables were treated using a two-stage mapping process that was constructed by mapping the relationship between unsigned decimal integers and discrete values. In this approach, GA was modified with a small generation gap that significantly reduced the function evaluations. However, the CI-SPF, CI-SAPF and CI-SAPF-CBO algorithms obtained better solutions compared to HPSO. In CI-SPF and CI-SAPF,

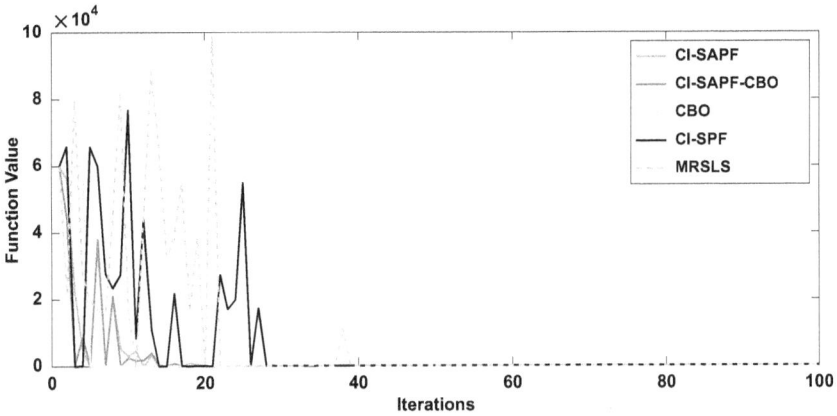

FIGURE 6.24 Comparison of the Convergence Curves of CI-SPF, MRSLS, CI-SAPF, CI-SAPF-CBO and CBO for Solving the Integer Linear Programming (Minimisation) Problem.

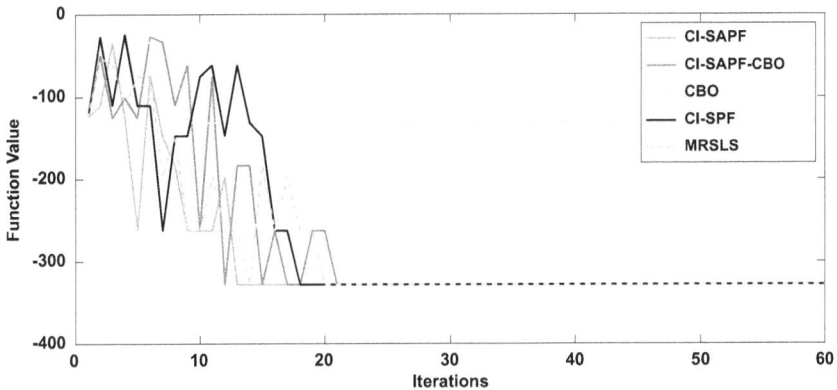

FIGURE 6.25 Comparison of the Convergence Curves of CI-SPF, MRSLS, CI-SAPF, CI-SAPF-CBO and CBO for Solving the Non-Convex Integer Programming (Formulation 1) Problem.

all the candidates are keen to improve their behaviour (function value), and a probability-based roulette wheel approach provides possible choices that enable them to follow the better solution, and CI further makes the solution move towards a global optimum. In CI-SAPF-CBO, the exploration quality of CBO controls the search space which makes the candidate achieve improved behaviour in the cohort.

The DHPSACO (Kaveh and Talatahari, 2009) and MBA (Sadollah et al., 2012) were employed with a modified feasibility-based (Deb,

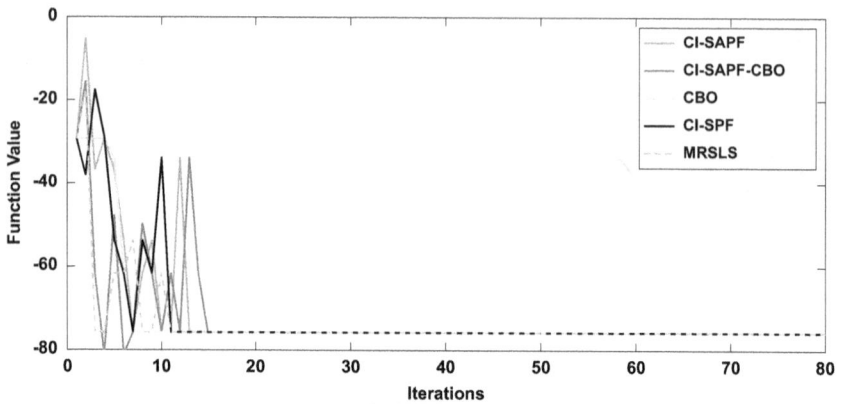

FIGURE 6.26 Comparison of the Convergence Curves of CI-SPF, MRSLS, CI-SAPF, CI-SAPF-CBO and CBO for Solving the Non-Convex Integer Programming (Formulation 2) Problem.

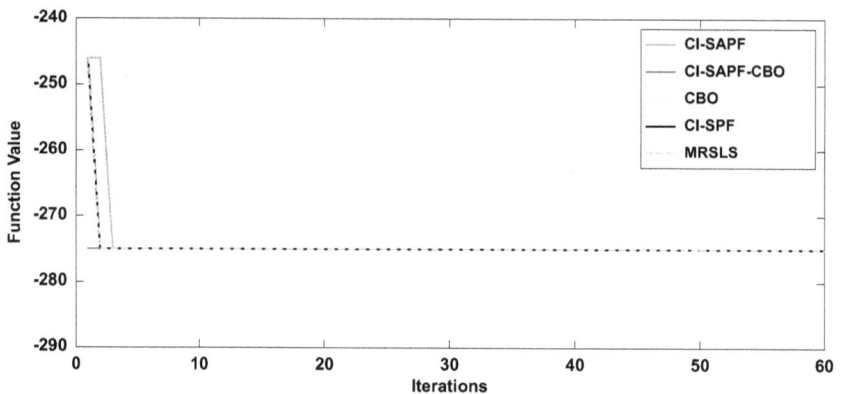

FIGURE 6.27 Comparison of the Convergence Curves of CI-SPF, MRSLS, CI-SAPF, CI-SAPF-CBO and CBO for Solving the Global Nonlinear Mixed Discrete Programming Problem.

2000) constraint handling approach and obtained better solutions with the least number of function evaluations. A harmony search strategy was adopted in DHPSACO to explore the search space that requires an additional parameter referred to as pitch adjustment rate to select the neighbourhood values. An SOS (Cheng and Prayogo, 2014) was also incorporated with a feasibility-based rule for solving design engineering problems such as the steeped beam for weight minimisation and the design of an I-section beam for minimum vertical deflection. A similar feasibility-based approach was

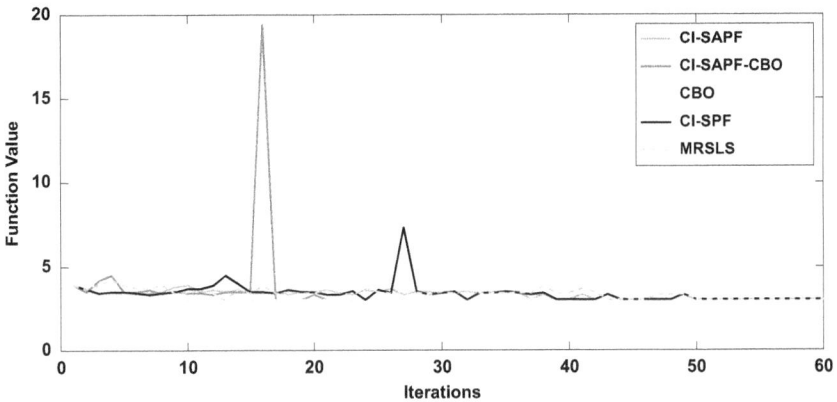

FIGURE 6.28 Comparison of the Convergence Curves of CI-SPF, MRSLS, CI-SAPF, CI-SAPF-CBO and CBO for Solving the 3-Bar Truss Structure Problem.

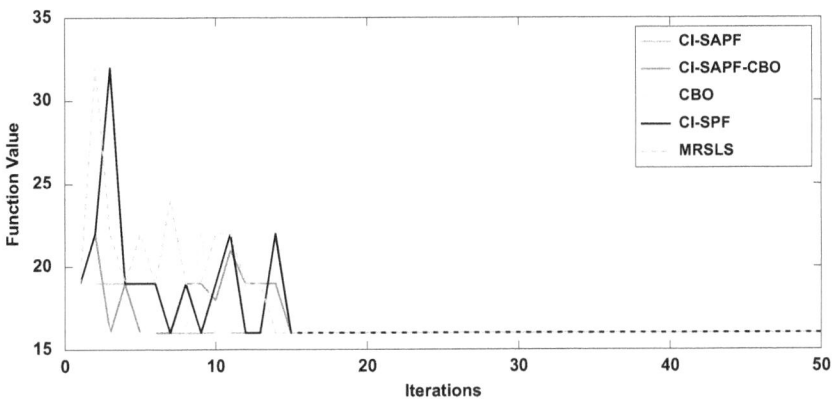

FIGURE 6.29 Comparison of the Convergence Curves of CI-SPF, MRSLS, CI-SAPF, CI-SAPF-CBO and CBO for Solving the Monotone (Function 1) Problem.

also adopted by Datta and Figueira (2011) in real-integer-discrete-coded PSO for solving the compound gear train design problem. Compared to the feasibility-based rule, CI-SAPF and CI-SAPF-CBO were observed to be superior in obtaining the same results as for the cantilever beam and compound gear train problems. For the I-section beam design problem, CI-SAPF and CI-SAPF-CBO obtained better solutions as compared to PSO and SOS. Moreover, the ADS algorithm (Hasançebi and Azad, 2015) is also compared with CI-SPF, CI-SAPF and CI-SAPF-CBO for the 10-bar case 1 truss structure problem. The ADS has achieved the best solution in 1000 function evaluations; however, an external penalty function was

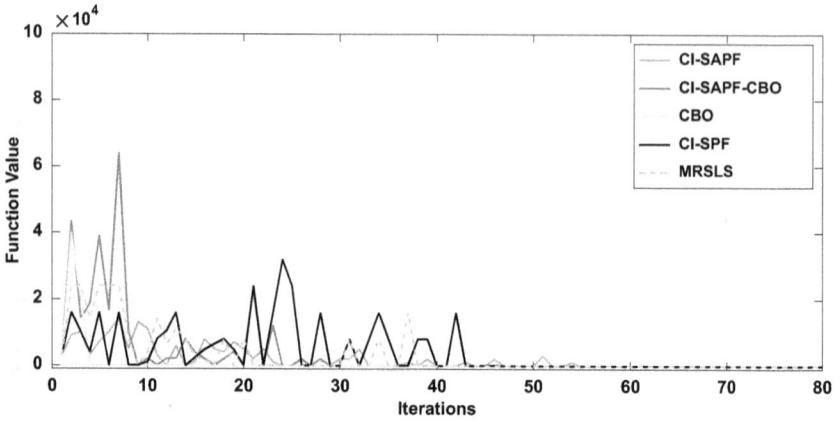

FIGURE 6.30 Comparison of the Convergence Curves of CI-SPF, MRSLS, CI-SAPF, CI-SAPF-CBO and CBO for Solving the Monotone (Function 2) Problem.

Note: * Infeasible solution.

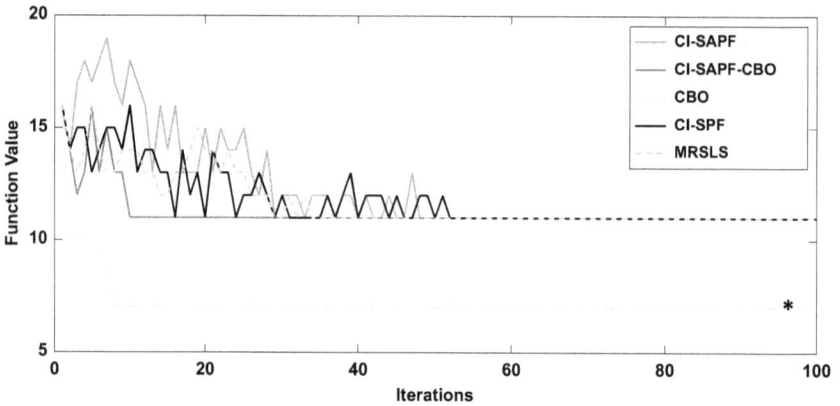

FIGURE 6.31 Comparison of the Convergence Curves of CI-SPF, MRSLS, CI-SAPF, CI-SAPF-CBO and CBO for Solving the Monotone (Function 3) Problem.

Note: * Infeasible solution.

incorporated for constraint handling in which an initial penalty coefficient needed to be set and further updated in every stagnation escape period. This required several preliminary trials and may increase the computational cost. The CI-SAPF and CI-SAPF-CBO are observed to be superior as they do not require penalty parameter tuning.

Thanedar and Vanderplaats (1995) proposed the stepped cantilever beam design problem. The B&B, specially developed approximation method and

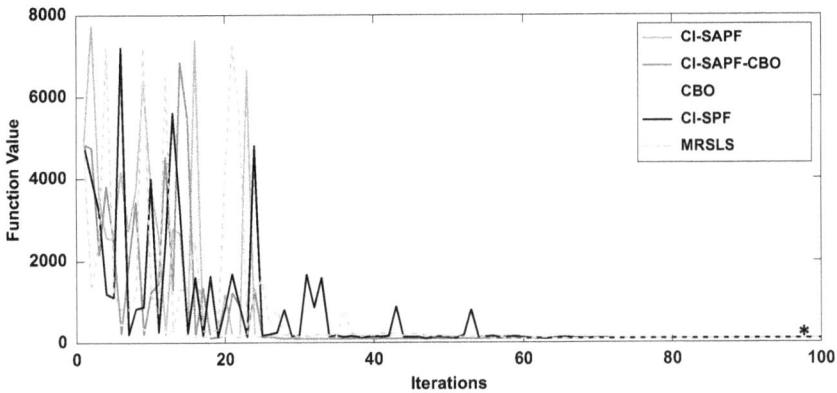

FIGURE 6.32 Comparison of the Convergence Curves of CI-SPF, MRSLS, CI-SAPF, CI-SAPF-CBO and CBO for Solving the Monotone (Function 4) Problem.

Note: * Infeasible solution.

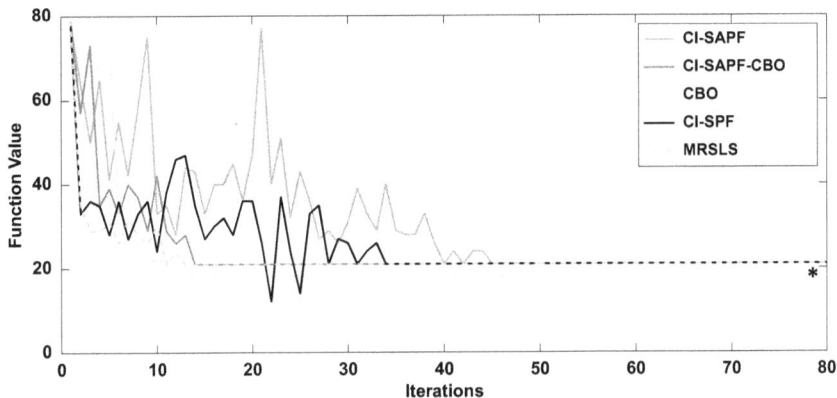

FIGURE 6.33 Comparison of the Convergence Curves of CI-SPF, MRSLS, CI-SAPF, CI-SAPF-CBO and CBO for Solving the Monotone (Function 5) Problem.

Note: * Infeasible solution.

ad-hoc methods such as SA and GA were described to solve the problem. The B&B method was observed to be a very expensive technique in terms of function evaluations and the approximation techniques failed to obtain a feasible solution. Thus, SA and GA were applied in order to solve discrete and mixed variable optimisation problems. Moreover, the penalty function was used to penalise the non-discrete variables, leading to an increase in terms of computational timing as well as function evaluations. The same penalty function approach is used in the CI algorithm to handle the

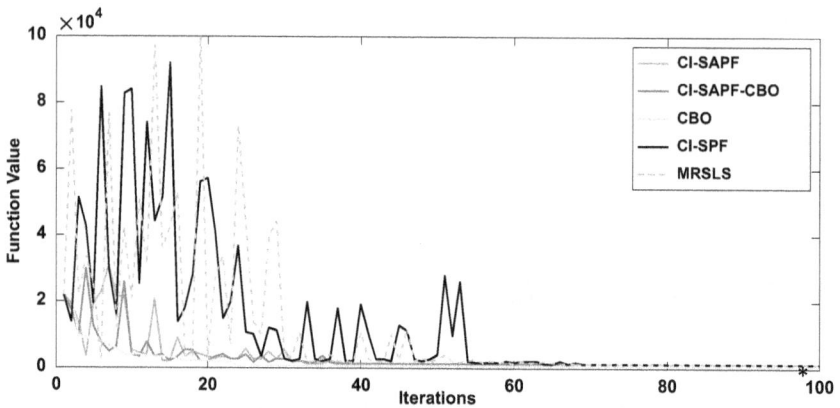

FIGURE 6.34 Comparison of the Convergence Curves of CI-SPF, MRSLS, CI-SAPF, CI-SAPF-CBO and CBO for Solving the Monotone (Function 6) Problem.

constraints; however, it did not increase computational cost significantly. CI-SPF is observed to be more efficient than Thanedar and Vanderplaats (1995) (see Table 3.15). Furthermore, CI-SAPF and CI-SAPF-CBO are used to solve the stepped beam cantilever beam design problem and obtained similar results as CI-SPF with less function evaluations.

The distributed optimisation technique of PC discussed in Kulkarni et al. (2016a) was modified to solve discrete and mixed variable truss and mechanical engineering problems. The PC algorithm obtained better results for solving 10-bar (Case 2), 25-bar (Case 1 and Case 2) and steeped cantilever beam problems as compared to CI-SPF, CI-SAPF and CI-SAPF-CBO (refer to Tables 6.2, 6.4, 6.5 and 6.16, respectively). The PC algorithm treats the complex problem/system in a distributed way which decomposes the entire system into a MAS. A feasibility-based rule was incorporated for constraint handling. The results obtained using PC were robust and exhibited better performance in terms of objective function than other techniques. However, due to its distributed nature, the convergence rate was slower which made it computationally inefficient. In CI-SPF, the self-supervising nature of the candidate that adopts the generic approach of accepting random behaviour to jump out of the local minima makes the CI-SPF, CI-SAPF and CI-SAPF-CBO algorithms computationally more efficient than PC.

It was noticed that the CI-SPF, CI-SAPF and CI-SAPF-CBO solutions are marginally worse than the PC solutions. Therefore, the comparative

TABLE 6.37 Comparison of the Function Values of CI-SPF and PC Associated to the Converged Function Evaluation of CI-SPF.

Test Examples	Converged Function Evaluation (FE) value for CI-SPF		CI-SPF Function Value		PC Function Value associated with FE value of CI-SPF	
	Best	Worst	Best	Worst	Best	Worst
10-bar Case 1	19250	11340	5490.7378	5508.3824	6163.3287	6376.0770
25-bar Case 1	21350	44275	516.2024	557.6885	564.1415	552.2852
and Case 2	14350	11200	500.0447	546.1536	735.0670	891.7324
38-bar	106666	96558	5900.3465	5938.0018	8120.3909	8138.7443
45-bar	163483	113400	14354.2722	14410.2122	21488.6204	24176.6442
72-bar Case 1	99288	153720	373.5427	391.6093	724.7563	718.4903
and Case 2	63000	123480	381.3418	395.1080	3046.7391	2875.2916
Reinforced Concrete Beam	22050	29820	359.2080	359.2080	359.2186	362.2537
Stepped beam problem	19740	11900	63893.4911	63946.7700	70185.6563	71698.4226
Speed reducer problem	16513	12495	2817.5581	2828.4442	3504.8167	5694.6172

Note: All the PC solutions found here are infeasible.

study of CI-SPF and PC solutions has been carried out in terms of function evaluations. The CI-SPF and PC algorithms are run 30 times when solving the test examples discussed here. As CI-SPF converged faster than PC, the function evaluations at which the CI-SPF solution converged are noted for every run. Similarly, the PC solution associated with the same number of function evaluations is also noted. Out of the 30 runs, the best and worst CI-SPF solutions and corresponding PC solutions with the same function evaluations are presented in Table 6.37. It is noted that the PC solutions are worse and infeasible. It is evident that the solutions obtained from CI-SPF were marginally worse.

A nonlinear integer and discrete programming method (NIDPM) (Sandgren, 1990) was proposed to solve mechanical design engineering problems such as the gear train design, the tension compression spring and the pressure vessel. A nonlinear B&B approach and exterior penalty function approach were incorporated to handle discrete and integer variables and to handle the constraints, respectively. The concept of the augmented Lagrange Multiplier (Kannan and Kramer, 1994) was incorporated with Powell's method and Fletcher and Reeves Conjugate Gradient method for solving the pressure vessel design problem. The zeroth order search (Powell's method) was found to be more efficient compared to the first order method. A dynamic constraint handling approach was incorporated with the augmented Lagrange Multiplier based method. With the same motivation of a parameter-less penalty function approach, He and Wang (2007) and Coello (2000) proposed the constraint handling approach as a co-evolutionary Particle Swarm Optimisation (CPSO) and self-adaptive penalty function using GA, respectively, for solving pressure vessel problems and the welded beam problem. For constraint handling, the penalty function approach was divided into two independent incremental weighing factors (He and Wang, 2007). From the results, it was noted that the function evaluations for GA (Coello, 2000) and CPSO (He and Wang, 2007) were 900,000 and 200,000, respectively. The CI-SAPF-CBO was observed to be more efficient with many less function evaluations (see Table 6.21) and with better function value than GA, CPSO and CI-SAPF. In the current work, CI-SAPF and CI-SAPF-CBO reduced the effort required to set the penalty parameter. This makes the algorithm superior to both other contemporary techniques and CI-SPF.

The discrete/integer variables linear programming problems (transportation problem, knapsack problem and linear integer programming) and nonlinear problems (dynamic problem, multistage problem and Rosen–Suzuki convex programming problem) were adopted from Srivastava and Fahim (2001) to validate the ability of CI-SPF, CI-SAPF and CI-SAPF-CBO. In Srivastava and Fahim (2001), a two-phase optimisation procedure was proposed in which a gradient-based steepest descent method for a feasible solution, and hem-stitching approach for an infeasible solution (when constraint violated) were distinctly incorporated. For the second phase, the integer combinations vector was obtained from the neighbourhood of the first phase. The process adopted in this work could make the algorithm more efficient and direct towards promising results. The CI-SPF, CI-SAPF and CI-SAPF-CBO algorithms are also applied to solve a Signomial Discrete Programming (SDP) problems (Tsai et al., 2002) (non-convex integer programming and global nonlinear mixed discrete programming problems). As compared to Floudas's approach (Tsai et al., 2002) for the transformation of the objective function and constraints (SDP problem) into a convex problem, CI-SPF, CI-SAPF and CI-SAPF-CBO are observed to be superior in obtaining a global solution.

REFERENCES

Arnout, S. (2011) 'International student competition in structural optimization', ISCSO, www.brightoptimizer.com

Bernardino, H.S., Barbosa, H.J.C. and Lemonge, A.C.C. (2007) 'A hybrid genetic algorithm for constrained optimization problems in mechanical engineering', *Proceedings of IEEE Congress Evolution Computations*, pp. 646–653.

Cheng, M.Y. and Prayogo, D. (2014) 'Symbiotic organisms search: A new metaheuristic optimization algorithm', *Computers and Structures*, Vol. 139, pp. 98–112.

Coello, C.A.C. (2000) 'Use of a self-adaptive penalty approach for engineering optimization problems', *Computer in Industry*, Vol. 41, pp. 113–127.

Coello, C.A.C. and Cortes, N.C. (2004) 'Hybridizing a genetic algorithm with an artificial immune system for global optimization', *Engineering Optimization*, Vol. 36, No. 5, pp. 607–634.

Coello, C.A.C. and Montes, E.M. (2002) 'Constraint-handling in genetic algorithms through the use of dominance-based tournament selection', *Advanced Engineering Informatics*, Vol. 16, No. 3, pp. 193–203.

Datta, D. and Figueira, J.R. (2011) 'A real-integer-discrete-coded particle swarm optimization for design problems', *Applied Soft Computing*, Vol. 11, pp. 3625–3633.

Deb, K. (1991) 'Optimal design of a welded beam via genetic algorithms', *AIAA J* 29, pp. 2013–2015.

Deb, K. (2000) 'An efficient constraint handling method for genetic algorithms', *Computer Methods in Applied Mechanics in Engineering*, Vol. 186, Nos. 2–4, pp. 311–338.

Deb, K. and Goyal, M. (1996) 'A combined genetic adaptive search (GeneAS) for engineering design', *Computer Science and Informatics*, Vol. 26, pp. 30–45.

Deb, K. and Srinivasan, A. (2005) 'Innovization: Innovation of design principles through optimization', KanGAL Report No. 2005007.

Eberhart, R. and Kennedy, J. (1995) 'A new optimizer using particle swarm theory'. *MHS'95. Proceedings of the Sixth International Symposium on micro Machine and Human Science*, Nagoya, pp. 39–43.

Efren, M., Coello, C.A.C. and Ricardo, L. (2003) 'Engineering optimization using a simple evolutionary algorithm', *Proceedings of 15th International Conference on Tools with Artificial Intelligence*, pp. 149–156.

Gandomi, A.H., Yang, X-S. and Alavi, A.H. (2011) 'Mixed variable structural optimization using firefly algorithm', *Computers and Structures*, Vol. 89, Nos. 23–24, pp. 2325–2336.

Gandomi, A.H., Yang, X.S. and Alavi, A.H. (2013) 'Cuckoo search algorithm: a metaheuristic approach to solve structural optimization problems', *Engineering with Computers*, Vol. 29, No. 1, pp. 17–35.

Hasançebi, O. and Azad, S.K. (2015) 'Adaptive dimensional search: A new metaheuristic algorithm for discrete truss sizing optimization', *Computers and Structures*, Vol. 154, pp. 1–16.

He, Q. and Wang, L. (2007) 'An effective co-evolutionary particle swarm optimization for constrained engineering design problem', *Engineering Applications of Artificial Intelligence*, Vol. 20, No. 1, pp. 89–99.

Kale, I.R. and Kulkarni, A.J. (2018) 'Cohort intelligence algorithm for discrete and mixed variable engineering problems', *International Journal of Parallel, Emergent and Distributed Systems*, Vol. 33, No. 6, pp. 627–662.

Kannan, B.K. and Kramer, S.N. (1994) 'An augmented Lagrange multiplier based method for mixed integer discrete continuous optimization and its applications to mechanical design', *Journal of Mechanical Design*, Vol. 116, No. 2, pp. 405–411.

Kaveh, A. and Mahdavi, V.R. (2014) 'Colliding bodies optimization: A novel meta-heuristic method', *Computers and Structures*, Vol. 39, 15 July, pp. 18–27.

Kaveh, A. and Mahdavi, V.R. (2015) *Colliding Bodies Optimization Extensions and Applications*, Springer.

Kaveh, A. and Talatahari, S. (2009) 'A particle swarm ant colony optimization for truss structures with discrete variables', *Journal of Constructional Steel Research*, Vol. 65, pp. 1558–1568.

Kulkarni, A.J., Kale, I.R. and Tai, K. (2013) 'Probability collectives for solving truss structure problems', *Proc. 10th World Congress on Structural and Multidisciplinary Optimization*.

Kulkarni, A.J., Kale I.R. and Tai K. (2016) 'Probability collectives for solving discrete and mixed variable problems', *International Journal of Computer Aided Engineering and Technology*, Vol. 8 No. 4, pp. 325–361.

Lawler, E.L. and Bell, M.D. (1966) 'A method for solving discrete optimization problem', *Operations Research*, Vol. 14, No. 6, pp. 1098–1112.

Lee, K.S., Geem, Z.W., Lee, S.H. and Bae, K.W. (2005) 'The harmony search heuristic algorithm for discrete structural optimization', *Engineering Optimization*, Vol. 37, No. 7, pp. 663–684.

Lemonge, A.C.C. and Barbosa, H.J.C. (2004) 'An adaptive penalty scheme for genetic algorithms in structural optimization', *International Journal of Numerical Methods and Engineering*, Vol. 59, No. 5, pp. 703–736.

Li, L.J., Huang, Z.B. and Liu, F. (2009) 'A heuristic particle swarm optimization method for truss structures with discrete variables', *Computers and Structures*, Vol. 87, Nos. 7–8, pp. 435–443.

Nanakorn, P. and Meesomklin, K. (2001) 'An adaptive penalty function in genetic algorithms for structural design optimization', *Computer and Structures*, Vol. 79, pp. 2527–2539.

Rajeev, S. and Krishnamoorthy, C.S. (1992) 'Discrete optimization of structures using genetic algorithm', *Journal of Structural Engineering*, Vol. 118, No. 5, pp. 1123–1250.

Rao, R.V. and Pawar, P.J. (2010) 'Parameter optimization of a multi-pass milling process using non-traditional optimization algorithms', *Applied Soft Computing*, Vol. 10, pp. 445–456.

Rao, R.V. and Savsani, V.J. (2012) *Mechanical Design Optimization Using Advanced Optimization Techniques*, Springer.

Rudolph, S. and Schmidt, J. (2012) 'International student competition in structural optimization', ISCSO, www.brightoptimizer.com

Sadollah, A., Bahreininejad, A., Eskandar, H. and Hamdi, M. (2012) 'Mine blast algorithm for optimization of truss structures with discrete variables', *Computers and Structures*, Vol. 49, No. 63, pp. 102–103.

Sandgren, E. (1990) 'Nonlinear integer and discrete programming in mechanical design optimization', *Journal of Mechanical Design*, Vol. 112, No. 2, pp. 223–229.

Shin, D.K., Gurdal, Z. and Griffin, O.H. (1990) 'A penalty approach for nonlinear optimization with discrete design variables', *Engineering Optimization*, Vol. 16, No. 1, pp. 29–42.

Sonmez, M. (2011) 'Artificial bee colony algorithm for optimization of truss structures', *Applied Soft Computing*, Vol. 11, No. 2, pp. 2406–2418.

Srivastava, V.K. and Fahim, A. (2001) 'A two-phase optimization procedure for integer programming problems', *Computers and Mathematics with Applications*, Vol. 42, pp. 1585–1595.

Thanedar, P.B. and Vanderplaats, G.N. (1995) 'Survey of discrete variable optimization for structural design', *Journal of Structural Engineering*, Vol. 121, No. 2, pp. 301–306.

Tsai, Jung-Fa, Li, Han-Lin and Hu, Nian-Ze (2002) 'Global optimization for signomial discrete programming problems in engineering design', *Engineering Optimization*, Vol. 34, No. 6, pp. 613–622.

Wu, S.J. and Chow, P.T. (1995) 'Steady-state genetic algorithms for discrete optimization of trusses', *Computers and Structures*, Vol. 56, No. 6, pp. 979–991.

Yun, Y.S. (2005) 'Study on Adaptive Hybrid Genetic Algorithm and its Applications to Engineering Design Problems', MSc thesis, Waseda University.

CHAPTER 7

Solution to
Real-World Applications

7.1 MULTI-PASS TURNING PROCESS PROBLEM

The multi-pass turning process problem is formulated based on the analysis given in Chen and Tsai (1996). The objective is to minimise the unit production cost by determining the optimal process parameters such as the cutting speed, feed rate, depth of cut and number of rough cuts. The production cost UC for the multi-pass turning operation is divided into four basic costs:

1. cutting cost by actual time in cut, C_M;
2. machine idle cost due to loading and unloading operations and idling tool motion, C_I;
3. tool replacement cost, C_R; and
4. tool cost, C_T.

For the unit production cost the objection function is defined as the sum of the cutting cost C_M, machine idling cost C_I, tool replacement cost C_R and tool cost C_T. The mathematical formulation in details is presented in Appendix.

DOI: 10.1201/9781003245193-7

The multi-pass turning process problem aims to minimise the unit production cost. The problem was formerly solved using an integer programming model (Gupta et al., 1995). In (Gupta et al., 1995), the multi-pass turning problem was modified to solve in two phases. In the first phase, the cost minimisation of the finish and rough passes was carried out, and in the second phase, the optimal combination of depth of cut was identified. Then Chen and Tsai (1996) proposed a nonlinear constrained optimisation algorithm comprising an SA and Hooke-Jeeves pattern search (SA/PS). This problem was also solved using GA, PSO, SA and Hybrid Genetic Simulated Swarm (HGSS) (Gayatri and Baskar, 2015).

The multi-pass turning process problem was formerly solved using GA, SA, PSO and HGGS in Gayatri and Baskar (2015). The constraint values were not illustrated in Gayatri and Baskar (2015), so the constraints are calculated based on those variables values that were available. It is noted that the cutting force and chip-tool interface temperature constraints are violated. This makes it evident that the solutions obtained using all four techniques are not feasible (see Table 7.1). Solving the same problem using the CI-SPF, CI-SAPF, CI-SAPF-CBO and CBO algorithms, the solutions are feasible with a best unit production cost of 2.5972 $ / *piece*. This solution is obtained using the CI-SAPF algorithm with average function evaluations of 1110 and an average computational time of 0.06 *sec*. The other statistical results obtained from CI-SPF, CI-SAPF and CI-SAPF-CBO are illustrated in Table 7.2. The CBO algorithm was not able achieve the feasible solution.

7.2 MULTI-PASS MILLING PROCESS PROBLEM

The multi-pass turning problem is adopted from Sonmez et al. (1999). This problem is solved to minimise the per unit production cost (i.e., to maximise the production rate). For the multi-pass milling process, the total production time T_{pr} comprises four elements of time as follows:

1. Machine preparation time (T_p)
2. Loading-Unloading time (T_L)
3. Process Adjusting and Quick return time (T_a)
4. Machining Time (T_m)
5. Tool changing time per component (T_c)

The mathematical formulation in details is presented in Appendix.

TABLE 7.1 Comparison of Results for the Multi-Pass Turning Process Problem.

Variables	GA (Gayatri and Baskar, 2015)	SA (Gayatri and Baskar, 2015)	PSO (Gayatri and Baskar, 2015)	HGSS (Gayatri and Baskar, 2015)	CI-SPF	CI-SAPF	CI-SAPF-CBO
V_r	389.15	447.94	499.99	499.9938	496.7222	495.0761	496.0539
f_r	0.7209	0.7255	0.8999	0.8939	0.1231	0.1163	0.107
d_r	2.03	2.21	2.50	2.50	2.3208	2.4306	2.5186
V_s	102.78	117.17	89.55	93.8986	132.5812	144.0232	167.1331
f_s	0.8788	0.5059	0.8999	0.8961	0.3324	0.3289	0.299
d_s	1.94	1.58	1.00	1.00	1.0492	1.0507	1.0644
g_1	115.5584	130.3326	188.2936	187.1010	-0.0000	0	-0.0000
g_2	-187.6149	-184.4716	-177.0964	-177.2108	-0.0195	-0.0195	-0.0195
g_3	-53639.0414	-65729.5709	-89846.4004	-89247.7831	-2.2283	-1.1584	-1.0325
g_4	400.4587	497.6636	691.2212	687.8356	-0.0024	-0.0091	-0.0105
g_5	-9.9195	-9.9733	-9.9156	-9.91635	-0.0010	-0.001	-0.0010
g_6	133.9725	50.0452	49.7859	49.4697	-0.0000	-0.0001	-0.0002
g_7	-196.3651	-197.7465	-198.2822	-198.2045	-0.0199	-0.0199	-0.0199
g_8	-17282.8387	-10585.6043	-7076.4803	-7760.8662	-0.7434	-0.7011	-0.6552
g_9	-131.5924	-241.1098	-227.9268	-214.1411	-0.0351	-0.0306	-0.0311
Cost \$ / piece	2.4205	3.0712	2.2255	2.2201	2.6108	2.5973	2.5992

TABLE 7.2   Statistical Results for the Multi-Pass Turning Process Problem Using CI-SPF, CI-SAPF and CI-SAPF-CBO.

Results	CI-SPF	CI-SAPF	CI-SAPF-CBO
Best	2.6108	2.5973	2.5992
Mean	2.6877	2.6909	2.7129
Worst	2.7901	2.8444	2.8361
Std. Dev.	0.0509	0.0685	0.0611
Avg. CPU Time	0.06	0.26	0.18
Avg. Fun. Eval.	506	1110	1624

The multi-pass milling process problem is solved using the CI-SPF, CI-SAPF, CI-SAPF-CBO and CBO algorithms. The problem contains mixed design variables i.e., feed rate ($f_z mm / tooth$) and spindle speed ($V\, m / min$) are continuous variables, and depth of cut (a) is a discrete variable. The total depth of cut is 5 mm. In the current work, the cutting strategy is adopted from Rao and Pawar (2010), i.e., if the number of passes is four, the first three are rough passes of $a = 1.5\,mm$ and for the fourth finish cut, $a = 0.5\,mm$. All the constraints considered here are formulated in $\leq$ form. The total production time obtained using CI-SPF, CI-SAPF and CI-SAPF-CBO is less as compared to ABC, PSO and SA (see Table 7.3). It is observed that the feed rate obtained using the proposed techniques is marginally higher than ABC, PSO and SA by satisfying all the constraints. The statistical results illustrated in Table 7.4 highlighted that CBO exhibited premature convergence. From the result comparison it can be seen that CI-SAPF-CBO has obtained a better solution than other contemporary techniques. The best production time noted is 2.15 min with average function evaluations of 1988 and average computational time of 12.26 sec.

7.3 CONCLUSION

The proposed CI-SPF, CI-SAPF and CI-SAPF-CBO algorithms have been successfully applied to solve real-world applications from the manufacturing domain. From the comparison it can be observed that CI-SAPF and CI-SAP-CBO obtain better solutions as compared to those from the GA, PSO, ABC and SA algorithms.

TABLE 7.3 Comparison of Results for the Multi-Pass Milling Process Problem.

Methods	Cutting strategy	f_z (mm/tooth)	V (m/min)	Arbor stress constraint	Arbor deflection constraint	Power constraint	T_{pr} (min)
ABC (Rao and Pawar, 2010)	$a_{rough} = 1.5$	0.337	46.982	4.708	435.02	0.004	3.24
	$a_{rough} = 1.5$	0.337	46.982	4.708	435.02	0.004	
	$a_{rough} = 1.5$	0.337	46.982	4.708	435.02	0.004	
	$a_{finish} = 0.5$	0.432	64.41	271.97	1.131	1.400	
PSO (Rao and Pawar, 2010)	$a_{rough} = 1.5$	0.34	46.61	1.5	431.9	0.01	3.24
	$a_{rough} = 1.5$	0.34	46.61	1.5	431.9	0.01	
	$a_{rough} = 1.5$	0.34	46.61	1.5	431.9	0.01	
	$a_{finish} = 0.5$	0.434	63.58	271.9	0.35	1.422	
SA (Rao and Pawar, 2010)	$a_{rough} = 1.5$	0.336	44.633	1.5	436.1	0.204	3.26
	$a_{rough} = 1.5$	0.336	44.633	1.5	436.1	0.204	
	$a_{rough} = 1.5$	0.336	44.633	1.5	436.1	0.204	
	$a_{finish} = 0.5$	0.429	57.23	273.91	2.296	1.683	
CBO	$a_{rough} = 1.5$	2.1204	33.6372	-3076.7856	-7298.4319	-3839.6497	2.23
	$a_{rough} = 1.5$	2.1204	33.6372	-3076.7856	-7298.4319	-3839.6497	
	$a_{rough} = 1.5$	2.1204	33.6372	-3076.7856	-7298.4319	-3839.6497	
	$a_{finish} = 0.5$	0.5144	31.4737	-4695.8656	-2031.3311	-3848.6419	

(continued)

TABLE 7.3 Comparison of Results for the Multi-Pass Milling Process Problem. (Continued)

Methods	Cutting strategy	fz (mm / tooth)	V (m / min)	Arbor stress constraint	Arbor deflection constraint	Power constraint	T_{pr} (min)
CI-SPF	$a_{rough} = 1.5$	0.9396	30.0011	-3911.8865	-8133.5328	-3844.8623	2.16
	$a_{rough} = 1.5$	0.9396	30.0011	-3911.8865	-8133.5328	-3844.8623	
	$a_{rough} = 1.5$	0.9396	30.0011	-3911.8865	-8133.5328	-3844.8623	
	$a_{finish} = 0.5$	0.6456	30.0256	-4648.9678	-1984.4333	-3848.4743	
CI-SAPF	$a_{rough} = 1.5$	0.9396	30.0021	-3911.8995	-8133.5458	-3844.8622	2.16
	$a_{rough} = 1.5$	0.9396	30.0021	-3911.8995	-8133.5458	-3844.8622	
	$a_{rough} = 1.5$	0.9396	30.0021	-3911.8995	-8133.5458	-3844.8622	
	$a_{finish} = 0.5$	0.8222	30.0686	-4589.8176	-19255.2832	-3848.1815	
CI-SAPF-CBO	$a_{rough} = 1.5$	0.6686	30.0012	-4139.5509	-8361.1971	-3845.9783	2.15
	$a_{rough} = 1.5$	0.6686	30.0012	-4139.5509	-8361.1971	-3845.9783	
	$a_{rough} = 1.5$	0.6686	30.0012	-4139.5509	-8361.1971	-3845.9783	
	$a_{finish} = 0.5$	0.5350	30.0406	-4688.3005	-2023.7660	-3848.6666	

TABLE 7.4 Statistical Results for the Multi-Pass Milling Process Problem Using CI-SPF, CI-SAPF, CI-SAPF-CBO and CBO.

Results	CI-SPF	CI-SAPF	CI-SAPF-CBO	CBO
Best	2.16	2.16	2.15	2.23
Mean	2.19	2.16	2.22	2.31
Worst	2.21	2.19	2.33	2.38
Std. Dev.	0.0192	0.0082	0.0468	0.0362
Avg. CPU Time	19.39	17.61	12.26	1.39
Avg. Fun. Eval.	3145	2856	1988	227

REFERENCES

Chen, M.C. and Tsai, D.M. (1996) 'A simulated annealing approach for optimization of multi-pass turning operations', *International Journal of Production Research*, Vol. 34, No. 10, pp. 2803–2825.

Gayatri, R. and Baskar, N. (2015) 'Evaluating process parameters of multi-pass turning process using hybrid genetic simulated swarm algorithm', *Journal of Advanced Manufacturing Systems*, Vol. 14, No. 4, pp. 215–233.

Gupta, R., Batra, J.L. and Lal, G.K. (1995) 'Determination of optimal subdivision of depth of cut in multipass turning with constraints', *International of Production and Research*, Vol. 33, No. 3, pp. 2555–2565.

Sonmez, A.I., Baykasoglu, A., Dereli, T. and Filiz, I˙.H. (1999) 'Dynamic optimization of multipass milling operations via geometric programming', *International Journal of Machine Tools and Manufacture*, Vol. 39, pp. 297–320.

CHAPTER **8**

Conclusions and Recommendations

8.1 CONCLUSIONS

A metaheuristic socio-based CI algorithm has been successfully applied to solve 10 discrete variable truss structure problems, 11 mixed variable mechanical engineering design problems and 17 linear and nonlinear integer variable test problems. The discrete variables associated with these problems were handled using a round off integer sampling approach. Initially, the CI algorithm was used with an SPF approach in order to handle the constraints. The SPF approach is simple and easy to apply to any constrained problems. It requires certain preliminary trials to set a penalty parameter that to penalises the infeasible function value for having violated constraints. The solutions obtained using CI-SPF were analysed and the results indicated that the CI approach was sufficiently robust with reasonable computational time period and fewer number of function evaluations as compared to other contemporary methods. In addition, a MRSLS method was also applied to solve all the problems considered in this work; however, the CI method outperformed the MRSLS method. This is due to the social behaviour of CI candidates that makes the candidates learn by following and interacting to improve their individual goal. Furthermore, this helps to evolve the entire cohort. Apart from these intrinsic worth properties, some disadvantages were also identified. It was observed that the quality of performance of CI was dependent on parameters such as the number of candidates C, the sampling space reduction factor R and the penalty parameter θ. The parameters were derived empirically over numerous experiments and their calibration required some preliminary trials. Additionally, it was observed that as the problem size increased, the computational time and function evaluations considerably increased.

In order to overcome the limitations of CI-SPF, the SAPF approach was developed for constraint handling. The penalty parameter required to run the SAPF approach was generated by the CI-SAPF algorithm itself which iteratively updated. The SAPF approach eliminated the need to set a penalty parameter. The behaviour of the pseudo objective function, penalty function and constraint violation were analysed based on the obtained value of the penalty parameter. The performance of the CI algorithm is essentially dependent on the sampling space reduction factor R which iteratively narrows down the sampling space resulting in better convergence of the function value. To set an appropriate value of R certain preliminary trials were required. In order to overcome this limitation, the CI-SAPF algorithm was hybridized with the CBO algorithm. This new algorithm, referred to as CI-SAPF-CBO, exhibited better performance with exploration and exploitation. Here, CI worked for the global search and CBO algorithm worked for the local search. The proposed CI-SAPF-CBO algorithm does not require any supplementary parameter to run the algorithm, thus reducing initial computational efforts.

The abilities of the CI-SAPF and CI-SAPF-CBO algorithms were tested by solving all 38 problems considered in this work. From the comparing the results from each algorithm, it was observed that the CI-SAPF algorithm performed exceptionally well due to it obtaining robust solutions with significantly less computational cost (i.e., computational time and function evaluations). The results from CI-SAPF-CBO were significantly superior and robust compared to those of CI-SAPF and other contemporary algorithms such as GA, PSO, FA, ACO, BA, CBO, SOS, etc. The proposed techniques were also applied to solve real world applications from manufacturing engineering and obtained better solutions compared to other contemporary techniques. The main objective of the hybridization of CI-SAPF-CBO was to make a generalised and parameter-less algorithm by removing the sampling space reduction factor R from CI-SAPF without degenerating the quality of solution. Here this objective is achieved by overcoming the limitations of the CI-SAPF algorithm with better quality of function value within a lesser number of function evaluations and computational time.

8.2 RECOMMENDATIONS

In the future, the proposed CI-SAPF and CI-SAPF-CBO algorithms could be applied to solve complex real-world applications from the combinatorial, manufacturing engineering domain, thermal engineering domain, healthcare domain, etc. The constraint handling SAPF approach can be incorporated in any of the nature-inspired optimisation techniques to improve its ability to handle linear and nonlinear constraints. The author is also interested in incorporating the characteristic of CI in other nature-inspired optimisation techniques to design new hybrid algorithms to solve complex engineering problems.

The proposed CI-SAPF and CI-SAPF-CBO algorithms could be used to reduce thermal pollution. This could be achieved by maximising the coefficient of performance (COP) of refrigeration systems by optimising or improving the heat transfer rate using solar and waste energy (Srivastava and Eames, 1998). The SAPF approach could be used to deal with the constraints such atmospheric conditions, properties of the absorber, properties of the refrigerant and the heat source. A round off integer sampling technique could be used to handle discrete and integer variables, where the variable could be input heat, mass of refrigerant, refrigeration effect, or specific cooling power, for example.

In the health care domain, the proposed CI-SAPF and CI-SAPF-CBO algorithms could be used to optimise the utilization of wards, scheduling of surgery, etc. This could be achieved by satisfying constraints such as the availability of the surgeon, daily and weekly specialty capacity and ensuring all blocks are allocated. The possible inputs could be number of surgeons, availability of beds in each ward, cases per block for the surgeon, etc.

The author is also interested in working in the finite element analysis domain. The CI has already been applied to a mesh smoothing problem (Sapre et al., 2018). The CI-SAPF and CI-SAPF-CBO could be modified in order to perform adaptive mesh refinement of complex geometry. This could be achieved by maintaining the convexity and geometry conditions. The CI-SPF, CI-SAPF and CI-SAPF-CBO were applied to solve two real world manufacturing problems (multi-pass milling and turning process problems).

REFERENCES

Sapre, M.S., Kulkarni, A.J., Chettiar, L., Deshpande, I. and Piprikar, B. (2018) 'Mesh smoothing of complex geometry using variations of cohort intelligence algorithm', *Evolutionary Intelligence*, pp. 1–16.

Srivastava, N.C. and Eames I.W. (1998) 'A review of adsorbents and adsorbates in solid vapour adsorption heat pump systems', *Applied Thermal Engineering*, Vol. 18, Nos. 9–10, pp. 707–714.

Problem Statements for the Truss Structure, Design Engineering, Linear and Nonlinear Programming and Manufacturing Problems

The problems considered here to examine the proposed CI-SAPF and CI-SAPF-CBO algorithms (see Chapters 4 and 5) have discrete, integer and mixed variables. There are 7 discrete variable truss structure problems, 11 mixed variable engineering design problems, 18 linear and nonlinear programming test problems and 2 real-world manufacturing problems. These problems have both linear and nonlinear constraints. The solutions and statistical analysis comparing with other contemporary techniques for these problems are illustrated in Chapters 3, 4, 5 and 6.

A.1 TRUSS STRUCTURE PROBLEMS

A.1.1 Problem 1: 6-Bar Truss Structure

The 6-bar truss structure (see Figure A.1) problem was formerly discussed in Nanakorn and Meesomklin (2001). There are six design variables (cross-sectional area) equal to the number of members of the truss. The cross-sectional area of each member is taken from the set of discrete values $A_i \in$ {1.62, 1.80, 1.99, 2.13, 2.38, 2.62, 2.63, 2.88, 2.93, 3.09, 3.13, 3.38, 3.47, 3.55, 3.63, 3.84, 3.87, 3.88, 4.18, 4.22, 4.49, 4.59, 4.80, 4.97, 5.12, 5.74, 7.22, 7.97, 11.50, 13.50, 13.90, 14.20, 15.50, 16.00, 16.90, 18.80, 19.90, 22.00, 22.90, 26.50, 30.00, 33.50} in^2 The objective function is to minimise the weight while satisfying the constraints for tension and compression stress in each member and deflection at every node of the truss structure. The allowable stress is given as $25000\,psi$ and allowable deflection is given as 2 in. The weight density of the material is $0.1\,lb/in^3$ and the modulus of elasticity is $10^7\,psi$. The problem definition for the truss structure is expressed in Eqs. 3.5, 3.6 and 3.7.

A.1.2 10-Bar Truss Structure

This discrete truss structure problem was previously discussed in Nanakorn and Meesomklin (2001) Li et al. (2009), Sonmez (2011) and Hasançebi and Azad (2015), and aims to minimise the weight W of an aluminium 2024-T3 10-bar truss structure as shown in Figure A.2 with material density $\rho\ 0.1\,lb/in^3$ and modulus of elasticity $E\ 10000\,ksi$. The maximum allowable stress σ_{max} in both tension and compression on every member i is $\pm25\,ksi$. The maximum allowable displacement u_{max} at every node in both

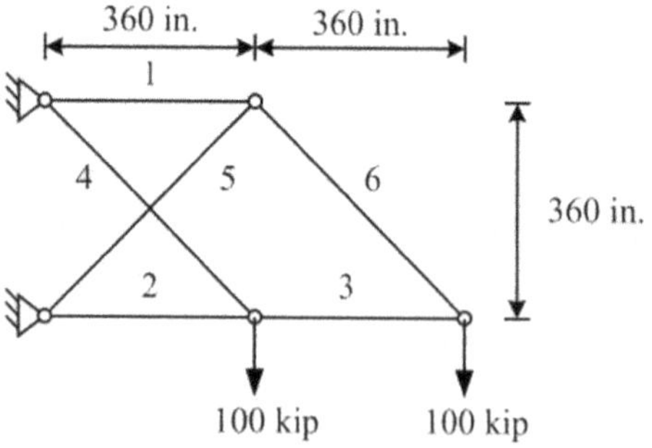

FIGURE A.1   Planer 6-Bar Truss Structure.

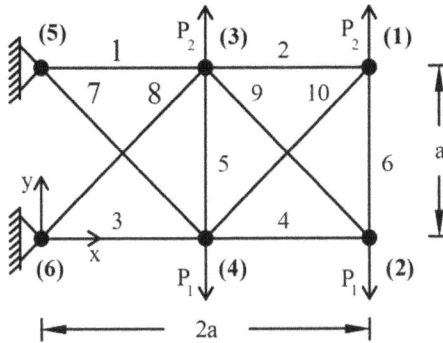

FIGURE A.2   Planer 10-Bar Truss Structure, a = 360 in^2.

the horizontal and vertical directions is ±2 in. The loading conditions are $P_1 = 100\,kips$ and $P_2 = 0$. There are ten design variables and two cases. For Case 1, the cross-sectional area of the discrete design variable should be selected from the set of $A_i \in$ {1.62, 1.80, 1.99, 2.13, 2.38, 2.62, 2.63, 2.88, 2.93, 3.09, 3.13, 3.38, 3.47, 3.55, 3.63, 3.84, 3.87, 3.88, 4.18, 4.22, 4.49, 4.59, 4.80, 4.97, 5.12, 5.74, 7.22, 7.97, 11.50, 13.50, 13.90, 14.20, 15.50, 16.00, 16.90, 18.80, 19.90, 22.00, 22.90, 26.50, 30.00, 33.50} in^2. For Case 2, $A_i \in$ {0.1,0.5,1.0,1.5,2.0, 2.5, 3.0, 3.5, 4.0, 4.5, 5.0, 5.5, 6.0, 6.5, 7.0, 7.5, 8.0, 8.5, 9.0, 9.5, 10.0, 10.5, 11.0, 11.5, 12.0, 12.5, 13.0, 13.5, 14.0, 14.5, 15.0, 15.5, 16.0, 16.5, 17.0, 17.5, 18.0, 18.5, 19.0, 19.5, 20.0, 20.5, 21.0, 21.5, 22.0, 22.5, 23.0, 23.5, 24.0, 24.5, 25.0, 25.5, 26.0, 26.5, 27.0, 27.5, 28.0, 28.5, 29.0, 29.5, 30.0, 30.5, 31.0, 31.5} in^2.

A.1.3 Spatial 25-Bar Truss Structure (Transmission Tower)

The Spatial 25-bar truss structure problem (Lee et al., 2005; Li et al., 2009; Kaveh and Talatahari, 2009; Kulkarni et al., 2013) aims to minimise the weight W of the 25-bar 3D spatial truss structure (see Figure A.3). The material weight density ρ is $0.1\,lb/in^3$ and the modulus of elasticity E is 10000 ksi. The maximum allowable stress σ_{max} in both tension and compression on every member i is ±40 ksi. The maximum allowable displacement u_{max} at every node in both horizontal and vertical directions is ±0.35 in. The structure is subjected to the two loading cases described in Table A.1. Design variable linking is employed to obtain a symmetric structure with respect to both the y-z and the x-z planes, and eight design variables are

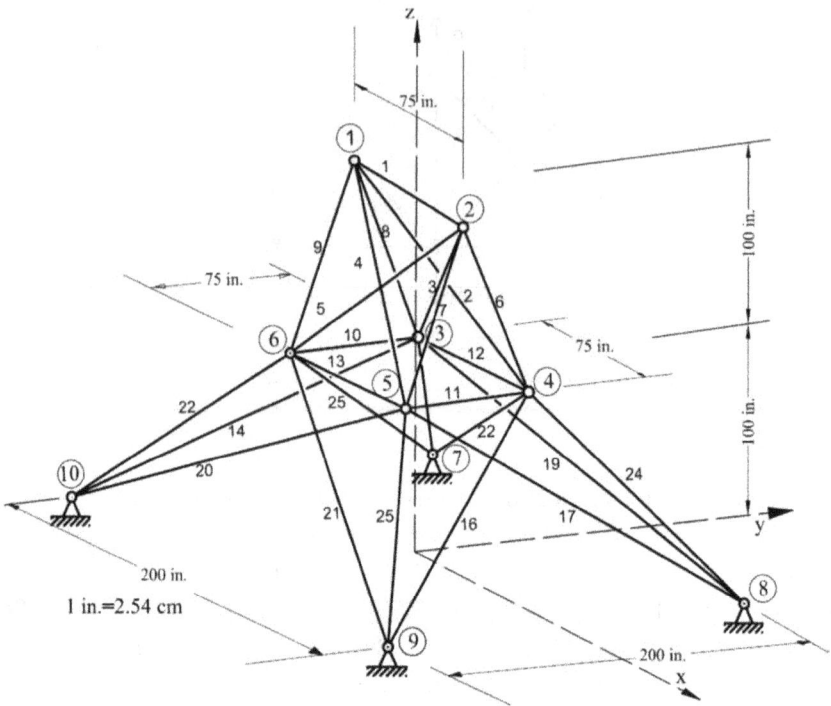

FIGURE A.3   Spatial 25-Bar Truss Structure.

TABLE A.1   Loading Conditions of Case 1 and Case 2 for the 25-Bar Truss Structure.

	Case 1			Case 2		
Nodes	$P_x(kips)$	$P_y(kips)$	$P_z(kips)$	$P_x(kips)$	$P_y(kips)$	$P_z(kips)$
1	0.0	20.0	−5.0	1.0	10.0	−5.0
2	0.0	−20.0	−5.0	0.0	10.0	−5.0
3	0.0	0.0	0.0	0.5	0.0	0.0
6	0.0	0.0	0.0	0.5	0.0	0.0

used to size the 25 members of the truss; this condition grouped the truss members as follows:

$$(1)A_1, (2)A_2 \sim A_5, (3)A_6 \sim A_9, (4)A_{10} \sim A_{11}, (5)A_{12} \sim A_{13},$$
$$(6)A_{14} \sim A_{17}, (7)A_{18} \sim A_{21}, (8)A_{22} \sim A_{25}$$

TABLE A.2   Available Cross-Sectional Area of the AISC Code.

No.	in^2	mm^2	No.	in^2	mm^2	No.	in^2	mm^2
1	0.111	71.613	22	2.380	1535.481	43	5.740	3703.218
2	0.141	90.968	23	2.620	1690.319	44	7.220	4658.055
3	0.196	126.451	24	2.630	1696.771	45	7.970	5141.925
4	0.250	161.290	25	2.880	1858.061	46	8.530	5503.215
5	0.307	198.064	26	2.930	1890.319	47	9.300	5999.988
6	0.391	252.258	27	3.090	1993.544	48	10.850	6999.986
7	0.442	285.161	28	3.130	2019.351	49	11.500	7419.340
8	0.563	363.225	29	3.380	2180.641	50	13.500	8709.660
9	0.602	388.386	30	3.470	2283.705	51	13.900	8967.724
10	0.766	494.193	31	3.550	2290.318	52	14.200	9161.272
11	0.785	506.451	32	3.630	2341.931	53	15.500	9999.980
12	0.994	641.289	33	3.840	2477.414	54	16.000	10322.560
13	1.000	645.160	34	3.870	2496.769	55	16.900	10903.204
14	1.228	792.256	35	3.880	2503.221	56	18.800	12129.008
15	1.266	816.773	36	4.180	2696.769	57	19.900	12838.684
16	1.457	939.998	37	4.220	2722.575	58	22.000	14193.520
17	1.563	1008.385	38	4.490	2896.768	59	22.900	14774.164
18	1.620	1045.159	39	4.590	2961.284	60	24.500	15806.420
19	1.800	1161.288	40	4.800	3096.768	61	26.500	17096.740
20	1.990	1283.868	41	4.970	3206.445	62	28.000	18064.480
21	2.130	1374.191	42	5.120	3303.219	63	30.000	19354.800
						64	33.500	21612.860

Case 1: The discrete variables are selected from the set $A_i \in$ (0.01, 0.4, 0.8, 1.2, 1.6, 2.0, 2.4, 2.8, 3.2, 3.6, 4.0, 4.4, 4.8, 5.2, 5.6, 6.0)

Case 2: The discrete variables are selected from the American Institute of Steel Construction (AISC) Code, listed in Table A.2.

A.1.4 Planer 38-Bar Truss Structure

This discrete truss structure problem was formerly discussed in Rudolph and Schmidt (2012) and Kulkarni et al. (2016) and aims to minimise the weight W of an AISI 1005 Steel 38-bar truss structure. The material weight density ρ is $0.283\,lb/in^3$ and the modulus of elasticity E is $30000\,ksi$. The maximum allowable stress σ_{max} in both tension and compression on every member i is $\pm 30\,ksi$. The maximum allowable displacement u_{max} at every node in both horizontal and vertical directions is $\pm 4\,in$. The load $P = 15\,kips$ is applied at the extreme end of the structure as illustrated in Figure A.4. The values of the cross-sectional area of every member should be selected from the discrete set $A_i \in \{0.1, 0.2, 0.3, \ldots, 14.8, 14.9, 15\}\,in^2$.

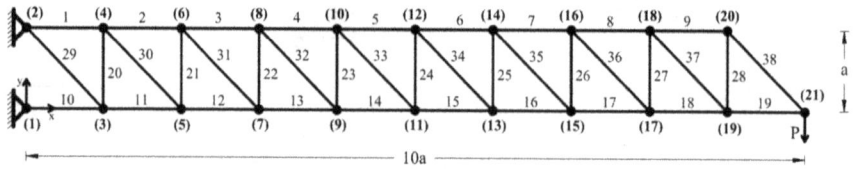

FIGURE A.4   Planer 38-Bar Truss Structure, a = 100 *in.*

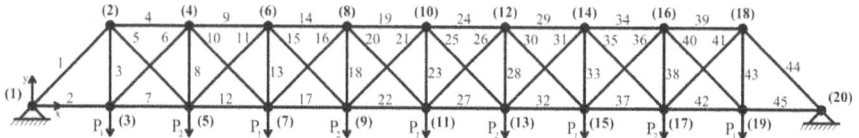

FIGURE A.5   Planer 45-Bar Truss Structure.

A.1.5 Planer 45-Bar Truss Structure

This 45-bar planer truss structure problem was previously described in Kulkarni et al. (2012) and Arnout (2011) and aims to minimise the weight W of the truss shown in Figure A.5. The symmetric truss structure members are divided into 23 groups and considered to be 23 sizing variables. They provide symmetry to the structure (see Figure A.5). The material weight density ρ of the structure is $0.283\,lb/in^3$ and the modulus of elasticity E is 30,000 ksi. The maximum allowable stress σ_{max} in both the tension and compression on every member i is ±30 ksi. The maximum allowable displacement u_{max} at every node in both horizontal and vertical directions is ±2 in. Nine vertical loads are applied simultaneously to the structure as follows:

Five lodes of $P_1 = 20\,kips$ are applied on nodes 3, 7, 11, 15 and 19
Four loads of $P_2 = 15\,kips$ are applied on nodes 5, 9, 13, and 17

The values of the cross-sectional area of every member should be selected from the discrete set $A_i \in \{0.1, 0.2, 0.3, ..., 14.8, 14.9, 15\}\,in^2$.

A.1.6 52-Bar Truss Structure

The members of this structure are divided into 12 groups: (1) A_1–A_4, (2) A_5–A_{10}, (3) A_{11}–A_{13}, (4) A_{14}–A_{17}, (5) A_{18}–A_{23}, (6)A_{24}–A_{26},(7) A_{27}–A_{30}, (8) A_{31}–A_{36}, (9) A_{37}–A_{39}, (10) A_{40}–A_{43}, (11) A_{44}–A_{49}, (12) A_{50}–A_{52}. The material density is $7860.0\,kg/m^3$ and the modulus of elasticity is $2.07 \times 105\,MPa$. The members are subjected to stress limitations of ±180 MPa (see Figure A.6). Loads of $Px = 100\,kN$ and $Py = 200\,kN$, are applied to the structure.

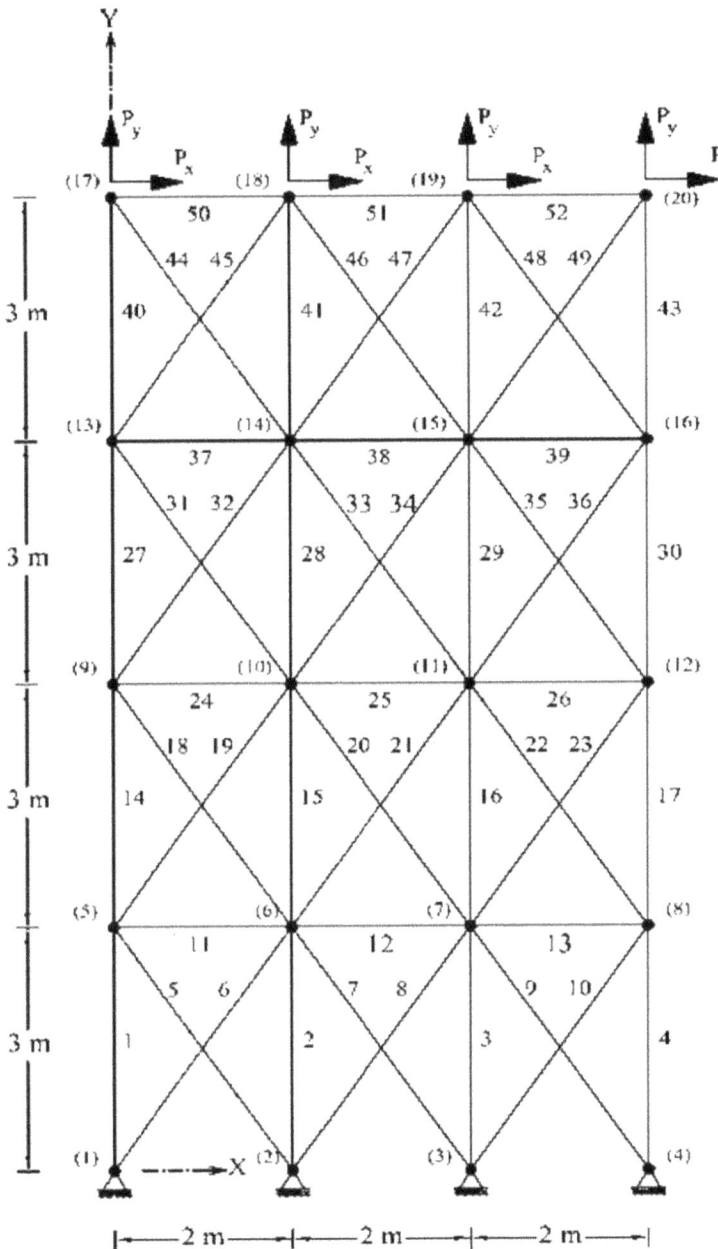

FIGURE A.6   52-Bar Truss Structure.

A.1.7 Spatial 72-Bar Truss Structure

For the 72-bar discrete spatial truss structure shown in Figure A.7 was previously solved in Wu and Chow (1995), Lee et al. (2005), Li et al. (2009),

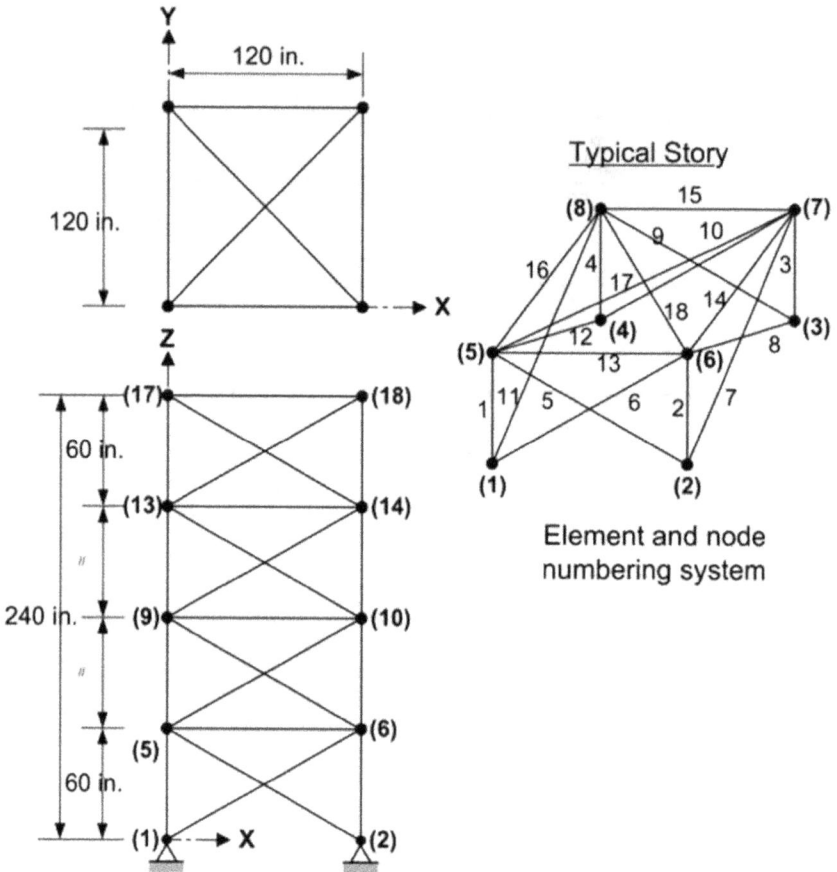

FIGURE A.7   Spatial 72-Bar Truss Structure.

Kaveh and Talatahari (2009) and Kulkarni et al. (2013) to minimise the weight W. The material density ρ is $0.1\,lb/in^3$ and the modulus of elasticity E is $10,000\,ksi$. The members are subjected to stress limits σ_{max} of $\pm25\,ksi$. The structure is subjected to load as listed in Table A.3. The nodes are subjected to displacement limits u_{max} of $\pm0.25\,in$. Design variable linking is employed to obtain a symmetric structure with respect to both the y-z and the x-z planes, and 16 design variables are used to size the 72 members of the truss. This condition groups the truss members as follows: (1) $A_1 \sim A_4$, (2) $A_5 \sim A_{12}$, (3) $A_{13} \sim A_{16}$, (4) $A_{17} \sim A_{18}$, (5) $A_{19} \sim A_{22}$, (6) $A_{23} \sim A_{30}$, (7) $A_{31} \sim A_{34}$, (8) $A_{35} \sim A_{36}$, (9) $A_{37} \sim A_{40}$, (10) $A_{41} \sim A_{48}$, (11) $A_{49} \sim A_{52}$, (12) $A_{53} \sim A_{54}$, (13) $A_{55} \sim A_{58}$, (14) $A_{59} \sim A_{66}$, (15) $A_{67} \sim A_{70}$, (16) $A_{71} \sim A_{72}$. Two optimisation cases are implemented.

TABLE A.3   Loading Condition for the 72-Bar Truss Structure.

| Nodes | Case 1 | | | Case 2 | | |
	$P_x(kips)$	$P_y(kips)$	$P_z(kips)$	$P_x(kips)$	$P_y(kips)$	$P_z(kips)$
17	5.0	5.0	−5.0	0.0	0.0	−5.0
18	0.0	0.0	0.0	0.0	0.0	−5.0
19	0.0	0.0	0.0	0.0	0.0	−5.0
20	0.0	0.0	0.0	0.0	0.0	−5.0

For Case (1): The cross-sectional area of design variable should be selected from the set $A_i \in \{0.1, 0.2, 0.3, 0.4, 0.5, 0.6, 0.7, 0.8, 0.9, 1.0, 1.1, 1.2, 1.3, 1.4, 1.5, 1.6, 1.7, 1.8, 1.9, 2.0, 2.1, 2.2, 2.3, 2.4, 2.5, 2.6, 2.7, 2.8, 2.9, 3.0, 3.1, 3.2\} \; in^2$. For Case (2): The cross-sectional area of design variable should be selected from Table A.2.

A.2 DESIGN ENGINEERING PROBLEMS

A.2.1 Stepped Cantilever Beam Design Problem

This mixed variable steeped beam problem was proposed by Thanedar and Vanderplaats (1995) and further discussed in Gandomi et al. (2011) and Kulkarni et al. (2016). This problem aims to minimise the volume V of the stepped cantilever beam of alloyed steel presented in Figure A.8. The problem formulation is as follows:

$$\text{Minimise } f = V = l\left(b_1 h_1 + b_2 h_2 + b_3 h_3 + b_4 h_4 + b_5 h_5\right) \qquad (A.1)$$

$$\text{subject to } g_1 = \frac{6Pl}{b_4 h_4^2} - \sigma_d \leq 0 \qquad (A.2)$$

$$g_2 = \frac{6P(2l)}{b_4 h_4^2} - \sigma_d \leq 0 \qquad (A.3)$$

$$g_3 = \frac{6P(3l)}{b_3 h_3^2} - \sigma_d \leq 0 \qquad (A.4)$$

$$g_4 = \frac{6P(4l)}{b_2 h_2^2} - \sigma_d \leq 0 \qquad (A.5)$$

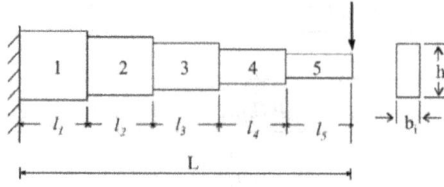

FIGURE A.8   Stepped Cantilever Beam.

$$g_5 = \frac{6P(5l)}{b_1 h_1^2} - \sigma_d \leq 0 \tag{A.6}$$

$$g_6 = \frac{Pl^3}{3E}\left(\frac{1}{I_5} + \frac{7}{I_4} + \frac{19}{I_3} + \frac{37}{I_2} + \frac{61}{I_1}\right) - D_{max} \leq 0 \tag{A.7}$$

$$g_7 = \frac{h_5}{b_5} - 20 \leq 0 \tag{A.8}$$

$$g_8 = \frac{h_4}{b_5} - 20 \leq 0 \tag{A.9}$$

$$g_9 = \frac{h_3}{b_3} - 20 \leq 0 \tag{A.10}$$

$$g_{10} = \frac{h_2}{b_2} - 20 \leq 0 \tag{A.11}$$

$$g_{11} = \frac{h_1}{b_1} - 20 \leq 0 \tag{A.12}$$

where b_1, b_2, b_3, h_1, h_2, h_3 are discrete variables and b_4, b_5, h_4, h_5 are continuous variables. The corresponding sampling sets and intervals are defined as follows:

$b_1 = \{1,2,3,4,5\};$ $b_2, b_2 = \{2.4,2.6,2.8,3.1\};$ $h_1, h_2 = \{45,50,55,60\};$
$h_3 = \{30,31,...,64,65\}; 1 \leq b_4, b_5 \leq 5$ and $30 \leq h_4, h_5 \leq 65.$

The values of constant terms are listed in Table A.4.

TABLE A.4 Constant Terms Provided for the Formulation of the Stepped Beam Design Problem.

Constants	Description	Values
P	Concentrated load	$1000\,lb$
σ_d	Design bending stress	$189e3\,psi$
E	Elastic modulus of the material	$30e6\,psi$
D_{max}	Allowable displacement	$11.5e6\,psi$
L	Total length of the five-stepped cantilever beam	$14\,in$

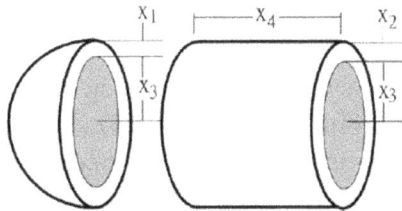

FIGURE A.9 Tube and Pressure Vessel.

A.2.2 Pressure Vessel Design Problems

Consider the optimal design problem of a pressure vessel given in Sandgren (1990) and depicted in Figure A.9 where x_1 (the spherical head thickness) and x_2 (the shell thickness) are discrete variables, and x_3 (the radius of the shell) and x_4 (the length of the shell) are continuous variables.

$$\text{Minimise } f(x) = 0.6224x_1x_3x_4 + 1.7781x_2x_3^2 + 3.1661x_1^2x_4 + 19.84x_1^2x_3 \tag{A.13}$$

$$\text{Subject to } g(x_1) = -x_1 + 0.0193x_3 \leq 0 \tag{A.14}$$

$$g(x_2) = -x_2 + 0.00954x_3 \leq 0 \tag{A.15}$$

$$g(x_3) = -\pi x_3^2 x_4 - \frac{4}{3}\pi x_3^3 + 750 \times 1728 \leq 0 \tag{A.16}$$

$$g(x_4) = -240 + x_4 \leq 0 \tag{A.17}$$

$$1 \leq x_1 \leq 1.375$$

$$0.625 \leq x_2 \leq 1$$

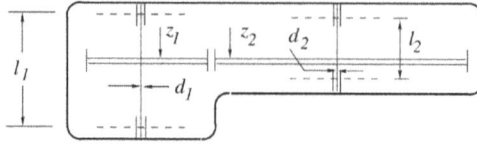

FIGURE A.10 Speed Reducer.

$$48 \leq x_3 \leq 52$$

$$90 \leq x_4 \leq 112$$

A.2.3 Speed Reducer Design Problem

This mixed variable problem was discussed in Bernardino et al. (2007), Coello and Cortes (2004), Efren et al. (2003) and Kulkarni et al. (2016) aims to minimise the weight f or W of the speed reducer of material 303 Stainless Steel presented in Figure A.10. The problem is formulated as follows:

$$\text{Minimise } f = W = 0.7854bz^2\left(3.3333\,Z^3 + 14.9334\,Z - 43.0934\right)$$
$$-1.508b\left(D_1^2 + D_2^2\right) + 7.4777\left(D_1^3 + D_2^3\right) \tag{A.18}$$
$$+0.7854\left(L_1 D_1^2 + L_2 D_2^2\right)$$

$$\text{Subject to } g_1 = \frac{27}{bz^2 Z} - 1 \leq 0 \tag{A.19}$$

$$g_2 = \frac{397.5}{bz^2 Z^2} - 1 \leq 0 \tag{A.20}$$

$$g_3 = \frac{1.93 D_1}{zZL_1^4} - 1 \leq 0 \tag{A.21}$$

$$g_4 = \frac{1.93 D_2^3}{zZL_2^4} - 1 \leq 0 \tag{A.22}$$

$$g_5 = \frac{\sqrt{\left(\dfrac{754 D_1}{zZ}\right)^2 + 16.9e6}}{110 L_1^3} - 1 \leq 0 \tag{A.23}$$

$$g_6 = \frac{\sqrt{\left(\dfrac{754D_2}{zZ}\right)^2 + 157.5e6}}{85L_2^3} - 1 \leq 0 \qquad \text{(A.24)}$$

$$g_7 = \frac{zZ}{40} - 1 \leq 0 \qquad \text{(A.25)}$$

$$g_8 = \frac{5z}{b} - 1 \leq 0 \qquad \text{(A.26)}$$

$$g_9 = \frac{b}{12z} - 1 \leq 0 \qquad \text{(A.27)}$$

$$g_{10} = \frac{1.5L_1 + 1.9}{D_1} - 1 \leq 0 \qquad \text{(A.28)}$$

$$g_{11} = \frac{1.1L_2 + 1.9}{D_2} - 1 \leq 0 \qquad \text{(A.29)}$$

where $2.6 \leq b \leq 3.6; 0.7 \leq z \leq 0.8; 7.3 \leq D_1 \leq 8.3; 7.8 \leq D_2 \leq 8.3; 2.9 \leq L_1 \leq 3.9;$ $5 \leq L_2 \leq 5.5$ and $17 \leq Z \leq 28$ are integer. The variables b, z, D_1, D_2, L_1, L_2 are continuous, and Z is a discrete variable.

A.2.4 Reinforced Concrete Beam Design

The mixed variable reinforced concrete beam problem (Gandomi et al., 2011; Shih and Yang, 2002; Montes and Ocana, 2009; Kulkarni et al., 2016) requires that beam design is achieved for minimum cost f. The beam is simply supported at both ends, spaced by 30 ft and is subjected to live load of 2000 lbf and dead load of 1000 lbf accounting for the beam weight as shown in Figure A.11. The problem formulation is as follows:

$$\text{Minimise } f = 29.4A + 0.6bh \qquad \text{(A.30)}$$

$$\text{subject to } g_1 = \frac{b}{h} - 4 \leq 0 \qquad \text{(A.31)}$$

$$g_2 = 180 + 7.375\frac{A^2}{h} - Ab \leq 0 \qquad \text{(A.32)}$$

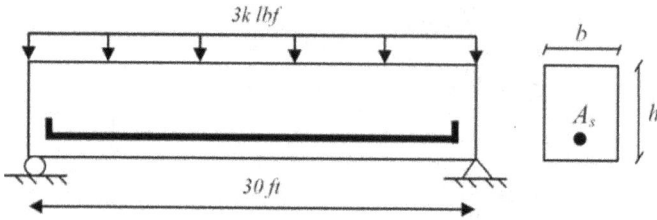

FIGURE A.11   Reinforced Concrete Beam.

$$M_u = 0.9A\sigma_y\left(0.8h\right)\left(1.0 - 5.59\frac{A\sigma_y}{0.8bh\sigma_c}\right) \geq 1.4M_d + 1.7M_t \quad \text{(A.33)}$$

where σ_c compressive strength $= 5\,ksi$

   σ_y yield strength $= 50\,ksi$

   A the cross-sectional area of the reinforced

   b the width of concrete beam

   h depth of the concrete beam

   M_u Flexural strength

   M_d Dead load moment $= 1350\ kip-in.$

   M_t Live load moment $= 2700\ kip-in.$

The width of the concrete beam b is an integer and must be selected from $\{28, 29, \ldots, 39, 40\}\ in$ whereas the effective depth is assumed to be $0.8\ h$ ($5 \leq h \leq 10$). The cross-sectional area A of the reinforced beam should be selected from the discrete values given in Table A.5.

A.2.5 Welded Beam Design Case 1

The welded beam design problem was discussed in Coello and Montes (2002), Coello (2000) and He and Wang (2007). Here, a welded beam is designed for minimum cost subject to constraints on the shear stress (τ), bending stress in the beam θ, buckling load on the bar (P_c), end deflection of the beam δ and side constraints. There are four design variables, as shown in Figure A.12, i.e., $h(x_1)$, $l(x_2)$, $t(x_2)$ and $b(x_2)$. The variable bounds are stated as $0.1 \leq x_1, x_2, x_3, x_4 \leq 10$.

TABLE A.5   Discrete Values of the Cross-Sectional Area A (in^2).

Bar Type	$A(in^2)$	Bar Type	$A(in^2)$	Bar Type	$A(in^2)$	Bar Type	$A(in^2)$
1#4	0.2	6#5	1.86	9#6	3.95	9#8	7.11
1#5	0.31	10#4, 2#9	2	4#9	3.96	12#7	7.2
2#4	0.4	7#5	2.17	13#5	4	13#7	7.8
1#6	0.44	11#4,5#6	2.2	7#7	4.03	10#8	7.9
3#4,1#7	0.6	3#8	2.37	14#5	4.2	8#9	8
2#5	0.62	12#4,4#7	2.4	10#6	4.34	14#7	8.4
1#8	0.79	8#5	2.48	15#5	4.4	11#8	8.69
4#4	0.8	13#4	2.6	6#8	4.65	15#7	9
2#6	0.88	6#6	2.64	8#7	4.74	12#8	9.48
3#5	0.93	9#5	2.79	11#6	4.8	13#8	10.27
5#4,1#9	1	14#4	2.8	5#9	4.84	11#9	11
6#4,2#7	1.2	15#4,5#7,3#9	3	12#6	5	14#8	11.06
4#5	1.24	7#6	3.08	9#7	5.28	15#8	11.85
3#6	1.32	10#5	3.1	7#8	5.4	12#9	12
7#4	1.4	4#8	3.16	13#8	5.53	13#9	13
5#5	1.55	11#5	3.41	10#7,6#9	5.72	14#9	14
2.8	1.58	8#6	3.52	14#6	6	15#9	15
8#4	1.6	6#7	3.6	8#8	6.16		
4#6	1.76	12#5	3.72	15#6,11#7	6.32		
9#4,3#7	1.8	5#8	3.82	7#9	6.6		

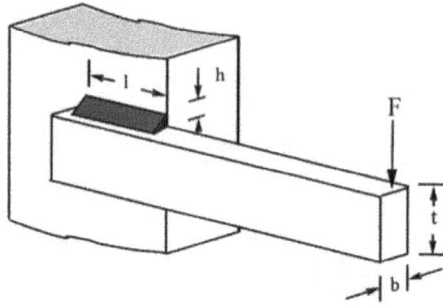

FIGURE A.12   Welded Beam Design Problem Case 1.

The problem can be mathematically formulated as follows:

$$\text{Minimise } f = 1.10471 x_1^2 x_2 + 0.04811 x_3 x_4 (14.0 + x_2) \tag{A.34}$$

$$\text{subject to } g_1 = \tau(x) - 13000 \leq 0 \tag{A.35}$$

$$g_2 = \sigma(x) - 30000 \leq 0 \tag{A.36}$$

$$g_3 = x_1 - x_4 \leq 0 \tag{A.37}$$

$$g_4 = 1.10471 x_1^2 x_2 + 0.04811 x_3 x_4 (14.0 + x_2) - 5.0 \leq 0 \tag{A.38}$$

$$g_5 = 0.125 - x_1 \leq 0 \tag{A.39}$$

$$g_6 = \delta(x) - 0.25 \leq 0 \tag{A.40}$$

$$g_7 = 6000 - P_c(x) \leq 0 \tag{A.41}$$

where

$$\tau(x) = \sqrt{(\tau')^2 + 2\tau'\tau'' \frac{x_2}{2R} + (\tau'')^2}$$

$$\tau' = \frac{6000}{\sqrt{2} x_1 x_2}$$

$$\tau'' = \frac{MR}{J}$$

$$M = 6000\left(14 + \frac{x_2}{2}\right)$$

$$J = 2\left\{\sqrt{2}x_1 x_2 \left[\frac{x_2^2}{12} + \left(\frac{x_1 + x_3}{2}\right)^2\right]\right\}$$

$$\sigma(x) = \frac{504000}{x_4 x_3^2}$$

$$\delta(x) = \frac{2.1952}{x_4 x_3^3}$$

$$P_c(x) = 64746.022\left(1 - 0.0282346x_3\right)x_3 x_4^3$$

A.2.6 Welded Beam Design Problem Case 2

In this problem, a rectangular beam is to be welded to a rigid support as a cantilever beam to carry a certain load without failure (Deb and Goyal (1996); Kennedy and Eberhart (1997)). There is an option for the beam to be welded either on two opposite sides or on all four sides of its rectangular cross-section. The beam may be of any of the four materials of steel, cast iron, aluminium and brass. It is necessary to design the system so as to minimise the overall construction cost. As shown in Figure A.13, apart from the two decision variables related to the choice of welding and beam material, the problem involves two pairs of design variables, each pair related to the dimensions of the weld and beam. The design variables related to the weld are its thickness (h) and length (l), and those related to the beam are its width (t) and thickness (b), where h, t and b are discrete variables in multiples of 0.0625 in. Accordingly, if the beam is to support

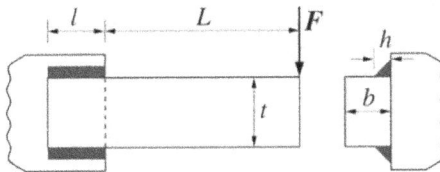

FIGURE A.13   Welded Beam Design Problem Case 2.

a constant load F at a fixed length L, the optimisation problem can be formulated as:

$$\text{Determine } x = (x_1, x_2, h, t, b, l)$$

$$\text{Minimise } f(x) = (1+c_1)(x_1 t + l)h^2 + c_2 tb(l+L) \tag{A.42}$$

$$\text{subject to } g_1 = \sigma(x) - S \le 0 \tag{A.43}$$

$$g_2 = F - P(x) \le 0 \tag{A.44}$$

$$g_3 = \delta(x) - \delta_{max} \le 0 \tag{A.45}$$

$$g_4 = \tau(x) - 0.577S \le 0 \tag{A.46}$$

$$x_1 \in \{0,1\}$$

$$x_2 \in \{0,1,2,3\}$$

$$h, t \text{ and } b \text{ are discrete}$$

$$h = b = [0.0625, 2.0] \ in; \ t = [2.0, \ 20.0] in; \ l = [1.0, 10.0] in$$

where

$$\sigma(x) = \frac{6FL}{bt^2}$$

$$P(x) = \frac{4.013bt^3\sqrt{EG}}{6L^2}\left[1 - \frac{t}{4L}\sqrt{\frac{E}{G}}\right]$$

$$\delta(x) = \frac{4FL^3}{Ebt^3}$$

$$\tau(x) = \sqrt{(\tau')^2 + (\tau'')^2 + 2\tau'\tau''\cos\theta}$$

$$\tau' = \frac{F}{A}$$

$$\tau'' = \frac{F(L+l/2)R}{J}$$

$$x_1 = 0: \begin{cases} A = \sqrt{2}hl \\ J = \sqrt{2}hl\left[\frac{(h+t)^2}{4} + \frac{l^2}{12}\right] \\ R = \frac{1}{2}\sqrt{l^2 + (h+t)^2} \\ \cos\theta = \frac{l}{2R} \end{cases}$$

$$x_1 = 1: \begin{cases} A = \sqrt{2}h(t+l) \\ J = \sqrt{2}hl\left[\frac{(h+t)^2}{4} + \frac{l^2}{12}\right] + \sqrt{2}hl\left[\frac{(h+l)^2}{4} + \frac{t^2}{12}\right] \\ R = \max\left\{\frac{1}{2}\sqrt{l^2+(h+t)^2}, \frac{1}{2}\sqrt{t^2+(h+l)^2}\right\} \\ \cos\theta = \begin{cases} \frac{l}{2R}, & \text{if } l < t \\ \frac{t}{2R}, & \text{otherwise} \end{cases} \end{cases}$$

Here, x_1 represents the type of weld, where $x_1 = 0$ is used for two-sided welding and $x_1 = 1$ is for four-sided welding. Similarly, x_2 indicates the type of material to be used for the beam, where $x_2 = 0$ means steel, $x_2 = 1$ means cast iron, $x_2 = 2$ means aluminum and $x_2 = 3$ means brass. The given values of F, L and δ_{max} are $6000\,lb$, $14\,in$ and $0.25\,in$, respectively. The values of the material properties S, E and G, and those of the cost terms c_1 and c_2 corresponding to different values of x_2, are given in Table A.6.

A.2.7 Multiple Disc Clutch Brake

This problem is taken from Deb and Srinivasan (2005). Figure A.14 shows a multiple disc clutch brake. The objective is to minimise the mass of the multiple disc clutch brake with five discrete variables: inner radius ($r_i = 60$,

TABLE A.6   Material Properties and Cost Terms for the Welded Beam Design Problem.

x_2	Material	S (kpsi)	E (Mpsi)	G (Mpsi)	c 1	c 2
0	Steel	30	30	12	0.1047	0.0481
1	Cast Iron	8	14	6	0.0489	0.0224
2	Aluminium	5	10	4	0.5235	0.2405
3	Brass	8	16	6	0.5584	0.2566

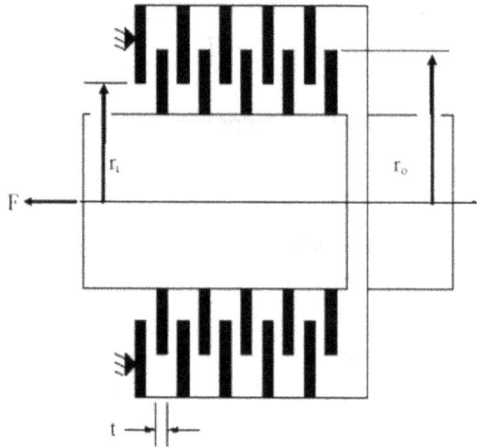

FIGURE A.14   Multiple Disc Clutch Brake.

61, 62 …,80), outer radius (r_o= 90, 91, 92 …110), thickness of discs ($t = 1$, 1.5, 2, 2.5, 3), actuating force ($F = 600, 610, 620, …, 1,000$) and number of friction surfaces ($Z = 2, 3, 4, 5, 6, 7, 8, 9$). The problem can be stated as:

$$\text{Minimise } f(x) = \pi\left(r_0^2 - r_i^2\right)t(Z+1)\rho \tag{A.47}$$

$$\text{subject to } g_1 = r_o - r_i - \Delta r \geq 0 \tag{A.48}$$

$$g_2 = l_{max} - (Z+1)(t+\delta) \geq 0 \tag{A.49}$$

$$g_3 = P_{max} - P_{rz} \geq 0 \tag{A.50}$$

$$g_4 = P_{max}V_{srmax} - P_{rz}V_{sr} \geq 0 \tag{A.51}$$

$$g_5 = V_{srmax} - V_{sr} \geq 0 \tag{A.52}$$

$$g_6 = T_{max} - T \geq 0 \tag{A.53}$$

A.2.8 Helical Compression Spring Design

The mixed variable problem of the helical compression spring design of alloyed steel (see Yun, 2005; Gandomi et al., 2011; Sandgren, 1990) presented in Figure A.15 aims to minimise the volume f (or V). The problem formulation is as follows:

$$\text{Minimise } f = V = \frac{\pi^2 Dd^2 (N+2)}{4} \tag{A.54}$$

$$\text{subject to } g_1 = \frac{8KP_{max}D}{\pi d^3} - S \leq 0 \tag{A.55}$$

$$g_2 = \left(\frac{P_{max}}{k} - 1.05(N+2)d\right) - L_{free} \leq 0 \tag{A.56}$$

$$g_3 = d_{min} - d \leq 0 \tag{A.57}$$

$$g_4 = (d+D) - D_{max} \leq 0 \tag{A.58}$$

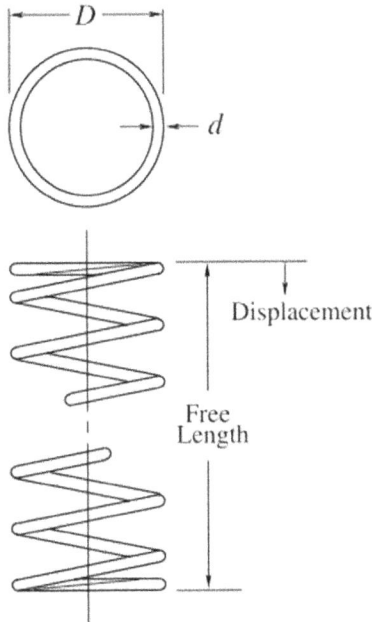

FIGURE A.15   Helical Compression Spring Design Problem.

TABLE A.7   Specific Design Size for the Wire Diameter d in^2.

0.0090	0.0150	0.0280	0.0720	0.1620	0.2830
0.0095	0.0162	0.0320	0.0800	0.1770	0.3070
0.0104	0.0173	0.0350	0.0920	0.1920	0.3310
0.0118	0.0180	0.0410	0.1050	0.2070	0.3620
0.0128	0.0200	0.0470	0.1200	0.2250	0.3940
0.0132	0.0230	0.0540	0.1350	0.2440	0.4375

$$g_5 = 3 - \frac{D}{d} \le 0 \tag{A.59}$$

$$g_6 = \delta_p - \delta_{pm} \le 0 \tag{A.60}$$

$$g_7 = \left(\frac{P_{max}}{k} - 1.05(N+2)d - L_f \right) - L_{free} \le 0 \tag{A.61}$$

$$g_8 = \delta_w - \left(\frac{P_{max} - P}{k} \right) \le 0 \tag{A.62}$$

where $C = \dfrac{D}{d}$, $K = \dfrac{4C-1}{4C-4} + \dfrac{0.615}{C}$, $k = \dfrac{Gd^4}{8ND^3}$, $\delta_p = \dfrac{P}{k}$, $1.0 \le D \le 30.0$ is continuous and $N \ge 1$ is integer. The values of the discrete variable d can be taken from those listed in Table A.7; the constant terms require to solve the problem are listed in Table A.8.

A.2.9 Minimise I-beam Vertical Deflection

The deflection minimisation of an I-beam using four variables was previously solved in Cheng and Prayogo (2014). Figure A.16 illustrates the goal of this Case – minimising the vertical deflection of an I-beam. The cross-sectional area and stress constraints must be satisfied simultaneously under given loads. The objective description of this study is to minimise the vertical deflection $f(x) = PL^3 / 48EI$ when beam length L and elasticity modulus E are, respectively, $5200\,cm$ and $523104\,kN/cm^2$. Thus, the objective function of the problem is considered to be:

$$\text{Minimise: } f(b,h,t_w,t_f) = \frac{5000}{\dfrac{t_w(h-2t_f)}{12} + \dfrac{bt_f^3}{6} + 2bt_f \left(\dfrac{h-t_f}{2} \right)^2} \tag{A.63}$$

TABLE A.8   The Associated Constant Terms Provided for the Formulation of the Helical Spring Design Problem.

Constant terms	Description	Values
P_{max}	Maximum work load	$1000\,lb$
S	Maximum shear stress	$189e3\,psi$
E	Elastic module of material	$30e6\,psi$
G	Shear module of material	$11.5e6\,psi$
L_{free}	Maximum coil free length	$14.0\,in$
d_{min}	Minimum wire diameter	$0.2\,in$
D_{max}	Maximum outside diameter of spring	$3.0\,in$
P	Preload compression force	$300.0\,lb$
δ_{pm}	Maximum deflection under preload	$6.0\,in$
δ_w	Deflection from preload position to maximum load position	$1.25\,in$

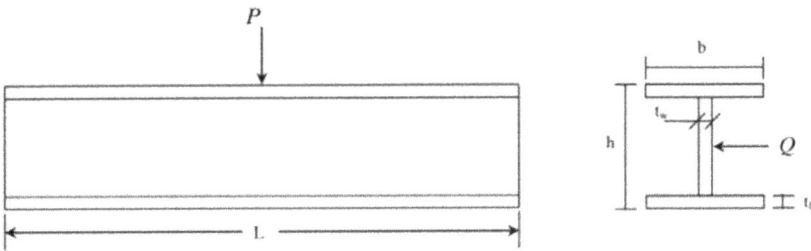

FIGURE A.16   I-Beam design.

subject to a cross-section area of less than $300\,cm^2$

$$g_1 = 2bt_w + t_w\left(h - 2t_f\right) \le 300 \qquad (A.64)$$

If the allowable bending stress of the beam is $56\,kN/cm^2$, the stress constraint is:

$$g_2 = \frac{18h \times 10^4}{ht_w\left(h - 2t_f\right)^3 + 2bt_w\left(4t_f^2 + 3h\left(h - 2t_f\right)\right)} + \frac{15b \times 10^3}{\left(h - 2t_f\right)t_w^3 + 2b^3 t_w} \le 56 \qquad (A.65)$$

Where the initial design spaces are $10 \le h \le 80$, $10 \le b \le 50$, $0.9 \le t_w \le 5$ and $0.9 \le t_f \le 5$.

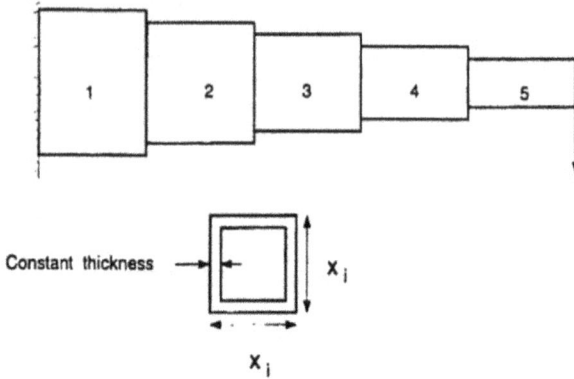

FIGURE A.17   Cantilever Beam.

A.2.10 Cantilever Beam

The cantilever beam problem was adopted from the work of Gandomi et al. (2011) and Cheng and Prayogo (2014). The cantilever beam shown in Figure A.17 comprises five elements. Each element has a hollow cross section of a fixed diameter. The beam is rigidly supported as shown, and a vertical force acts at the free end of the cantilever. The problem presented is to minimise the beam weight. The design variable is the height (or width) x_i of each beam element. Bound constraints are set as $0.01 \leq x_i \leq 100$. The problem is formulated using classical beam theory as:

$$\text{Minimise } f(x) = 0.0624(x_1 + x_2 + x_3 + x_4 + x_5) \tag{A.66}$$

$$\text{Subject to } g1 = \frac{61}{x_1^3} + \frac{37}{x_2^3} + \frac{61}{x_3^3} + \frac{61}{x_4^3} + \frac{61}{x_5^3} \leq 1 \tag{A.67}$$

A.2.11 The Gear Train Design Problem

This problem involves the optimisation of the gear ratio of a compound gear train, so that a desired motion/power can be transmitted from one shaft to another. As shown in Figure A.18, the example gear train arrangement consists of two pairs of gearwheels, $a - c$ and $b - d$, where a and b are driving gearwheels, and c and d are driven gearwheels. The overall gear ratio of the arrangement can be expressed as:

$$\text{Gear Ratio} = \frac{\Pi(\text{Number of teeth on Driving Gear Wheel})}{\Pi(\text{Number of teeth on Driven Gear Wheel})} = \frac{z_a z_b}{z_c z_d} \tag{A.68}$$

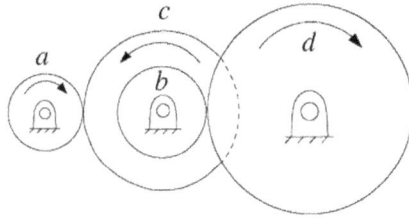

FIGURE A.18   Gear Train.

where z_a, z_b, z_c and z_d denote the numbers of teeth on the gears a, b, c and d respectively. It is necessary to determine the values of z_a, z_b, z_c and z_d so that a gear ratio as close as possible to 1/6.931, can be obtained. The only constraint in the problem is that the number of teeth on any gear should be in the range $\{12,13,\ldots,59,60\}$. Accordingly, the optimisation problem can be formulated as follows:

$$\text{Minimise } f(x) = \left(\frac{1}{6.931} - \frac{z_a z_b}{z_c z_d} \right)^2 \tag{A.69}$$

$$\text{subject to } g(x) = 12 \le z_a, z_b, z_c, z_d \le 60 \tag{A.70}$$

$$z_a, z_b, z_c \text{ and } z_d \text{ are integers}$$

A.3 TEST FUNCTIONS

A.3.1 Problem 3: Integer Linear Programming

$$\text{Maximise } f(x) = 3x_1 + x_2 + 2x_3 + x_4 - x_5 \tag{A.71}$$

$$\text{subject to } g1 = 25x_1 - 40x_2 + 16x_3 + 21x_4 + x_5 \le 300 \tag{A.72}$$

$$g2 = x_1 + 20x_2 - 50x_3 + x_4 - x_5 \le 200 \tag{A.73}$$

$$g3 = 60x_1 + x_2 - x_3 + 2x_4 + x_5 \le 600 \tag{A.74}$$

$$g4 = -5x_1 + 4x_2 + 15x_3 - x_4 + 65x_5 \le 700 \tag{A.75}$$

$$0 \le x_i \le 200, \ i = 1,\ldots,5 \text{ (integer)}$$

A.3.2 Problem 4: Rosen–Suzuki Test Problem Convex Programming Problem

$$\text{Minimise } f(x) = x_1^2 + x_2^2 + 2x_3^2 + x_4^2 - 5x_1 - 5x_2 - 21x_3 + 7x_4 \quad \text{(A.76)}$$

$$\text{Subject to } g1 = \sum_{i=1}^{4} x_i^2 + \sum_{i=1}^{3} (-1)^i x_{1+1} \le 8 \quad \text{(A.77)}$$

$$g1 = (x_1 - 1) + 2x_2^2 + x_3^2 + x_4 (2x_4 - 1) \le 10 \quad \text{(A.78)}$$

$$g1 = 2x_1 (x_1 + 1) + x_2 (x_2 - 1) + x_3^2 - x_4 \le 5 \quad \text{(A.79)}$$

$$-10 \le x_i \le 20, \ i = 1,\dots,4 \ (\text{integer})$$

A.3.3 Dynamic Variable Problem

$$\text{Maximise } f(x) = x_1^2 - 2x_1 + x_2^2 + 3x_2 + x_3^2 + 4x_3 \quad \text{(A.80)}$$

$$\text{subject to } g1 = x_1^2 + 2x_2^2 + 3x_3^2 \le 17 \quad \text{(A.81)}$$

$$0 \le x_i \le 10, \ i = 1,\dots,3 \ (\text{integer})$$

A.3.4 Transportation Problem

$$\text{Minimise } f(x) = 0.9x_1 + x_2 + x_3 + x_4 + 1.4x_5 \\ + 0.8x_6 + 1.3x_7 + x_8 + 0.8x_9 \quad \text{(A.82)}$$

$$\text{subject to } g1 = x_1 + x_2 + x_3 \le 20 \quad \text{(A.83)}$$

$$g2 = x_4 + x_5 + x_6 \le 15 \quad \text{(A.84)}$$

$$g3 = x_7 + x_8 + x_9 \le 10 \quad \text{(A.85)}$$

$$g4 = x_1 + x_4 + x_7 \ge 5 \quad \text{(A.86)}$$

$$g5 = x_2 + x_5 + x_8 \ge 20 \quad \text{(A.87)}$$

$$g6 = x_3 + x_6 + x_9 \ge 20 \quad \text{(A.88)}$$

$$0 \le x_i \le 100, \ i = 1,\dots,9 \ (\text{integer})$$

A.3.5 Multistage Problem

$$\text{Maximise } f(x) = 13x_1 - 5x_2^2 + 30.2x_2 - x_1^2 + 10x_3 + 2.5x_3^2 \quad \text{(A.89)}$$

$$\text{subject to } g1 = 2x_1 + 4x_2 + 5x_3 \le 10 \quad \text{(A.90)}$$

$$g2 = x_1 + x_2 + x_3 \le 5 \quad \text{(A.91)}$$

$$x_i \le 0, \quad i = 1,2,3$$

A.3.6 Knapsack Problem

$$\text{Maximise } f(x) = 592x_1 + 381x_2 + 273x_3 + 55x_4 + 48x_5 + 37x_6 + 23x_7 \text{ (A.92)}$$

$$\text{subject to } g1 = 3534x_1 + 2345x_2 + 1767x_3 + 589x_4 + 528x_5 \\ + 451x_6 + 304x_7 \le 119567 \quad \text{(A.93)}$$

$$0 \le x_i \le 100, \quad i = 1,\ldots,7 \text{ (integer)}$$

A.3.7 Integer Linear Programming

$$\text{Maximise } f(x) = 3x_1 + 4x_2 \quad \text{(A.94)}$$

$$\text{subject to } g1 = 3x_1 - x_2 \le 12 \quad \text{(A.95)}$$

$$g2 = 7x_1 + 11x_2 \le 88 \quad \text{(A.96)}$$

$$0 \le x_i \le 100, \quad i = 1,\ldots,2 \text{ (Integer)}$$

A.3.8 Non-Convex Integer Problem

The following non-convex minimisation problem containing integer variables has previously been solved by Tsai et al. (2002):

$$\text{Minimise } f(x) = x_1^2 x_2^{3.5} x_3 - x_2 x_3^{2.6} - x_1^3 \quad \text{(A.97)}$$

$$\text{subject to } g(x_1) = x_1 + x_2 + x_3 \le 10 \quad \text{(A.98)}$$

This problem was solved using non-convex integer programming as well as the relaxation method. For both techniques, the sampling interval was selected as follows:

Formulation 1 (non-convex integer programming): $1 \le x_1 \le 5; 1 \le x_2 \le 5;$
$1 \le x_3 \le 5$

Formulation 2 (relaxation method): $0 \le x_1 \le 5; 0 \le x_2 \le 5; 0 \le x_3 \le 5$

x_1, x_2, x_3 are integer variables.

A.3.9 Global Nonlinear Programming

$$\text{Minimise } f(x) = 2x_1^2 + x_2^2 - 16x_1 x_2 - 10x_2 \quad (A.99)$$

Subject to

$$g1 = \left(x_1^2 - 6x_1 + 4x_2 - 11\right)\left[\left(3.25x_1 - 3.1x_2\right)^2 + \left(x_1 + x_2 - 6.35\right)^2\right]$$
$$\left[\left(3.55x_1 - 3.3x_2\right)^2 + \left(x_1 + x_2 - 6.85\right)^2\right]\left[\left(3.6x_1 - 3.5x_2\right)^2 + \left(x_1 + x_2 - 7.1\right)^2\right]$$
$$\left[\left(3.8x_1 - 4.1x_2\right)^2 + \left(x_1 + x_2 - 7.9\right)^2\right]^2 = 0 \quad (A.100)$$

$$g2 = -x_1 x_2 + 3x_2 + e^{x_1 - 3} - 1 \le 0 \quad (A.101)$$

$$3 \le x_1 \le 6$$

$$3 \le x_2 \le 5$$

where x_1 is an integer variable and x_2 is a discrete variable in multiples of 0.2.

A.3.10 3-Bar Truss Structure

$$\text{Minimise } f(x) = 2x_1 + x_2 + \sqrt{2x_3} \quad (A.102)$$

$$\text{subject to } g_1 = 1 - \frac{\sqrt{3}x_2 + 1.932x_3}{1.5x_1 x_2 + \sqrt{2}x_2 x_3 + 1.319x_3} \ge 0 \quad (A.103)$$

$$g_2 = 1 - \frac{0.634x_1 + 2.828x_3}{1.5x_1 x_2 + \sqrt{2}x_2 x_3 + 1.319x_3} \ge 0 \quad (A.104)$$

$$g_3 = 1 - \frac{0.5x_1 + 2x_2}{1.5x_1x_2 + \sqrt{2}x_2x_3 + 1.319x_3} \geq 0 \qquad \text{(A.105)}$$

$$g_4 = 1 + \frac{0.5x_1 - 2x_2}{1.5x_1x_2 + \sqrt{2}x_2x_3 + 1.319x_3} \geq 0 \qquad \text{(A.106)}$$

$$x_i = \{0.1, 0.2,\ 0.3,\ 0.5,\ 0.8,\ 1.0,\ 1.21\},\ i = 1, 2, 3$$

A.3.11 Monotone Objective Functions

The series of six problems have contrived constraints and mostly non-linear objective functions that consist of discrete (integer) variables. These problems were solved by transforming the integer variable into a binary variable, putting the integer linear programming problem into a different monotone function with the appropriate grouping of positive and negative coefficients.

A.3.11.1 Problem 1

$$\text{Minimise } f(x) = x_1^2 + x_2^2 + x_3^2 + x_4^2 + x_5^2 \qquad \text{(A.107)}$$

$$\text{subject to } g(x_1) = x_1 + 2x_2 + x_4 \geq 4 \qquad \text{(A.108)}$$

$$g(x_2) = x_2 + 2x_3 \geq 3 \qquad \text{(A.109)}$$

$$g(x_3) = x_1 + 2x_5 \geq 5 \qquad \text{(A.110)}$$

$$g(x_4) = x_1 + x_2 + 2x_3 \geq 6 \qquad \text{(A.111)}$$

$$g(x_5) = 2x_1 + x_3 \leq 4 \qquad \text{(A.112)}$$

$$g(x_6) = x_1 + 4x_5 \geq 13 \qquad \text{(A.113)}$$

$$g(x_7) = x_j \leq 3 (j = 1, 2, \ldots 5) \qquad \text{(A.114)}$$

$$x_j \geq 0 \text{ and integer} (j = 1, 2, \ldots 5)$$

A.3.11.2 Problem 2

$$\text{Minimise } f(x) = x_1 x_7 + 3x_2 x_6 + x_3 x_5 + 7x_4 \tag{A.115}$$

$$\text{subject to } g(x_1) = x_1 + x_2 + x_3 \geq 6 \tag{A.116}$$

$$g(x_2) = x_4 + x_5 + 6x_6 \geq 8 \tag{A.117}$$

$$g(x_3) = x_1 x_6 + x_2 + 3x_5 \geq 7 \tag{A.118}$$

$$g(x_4) = 4x_2 x_7 + 3x_4 x_5 \geq 25 \tag{A.119}$$

$$g(x_5) = 3x_1 + 2x_3 + x_5 \geq 7 \tag{A.120}$$

$$g(x_6) = 4x_1 + 2x_3 + x_6 x_7 \leq 15 \tag{A.121}$$

$$g(x_7) = x_4 \leq 15 \tag{A.122}$$

$$g(x_8) = x_5 \leq 15 \tag{A.123}$$

$$g(x_9) = x_j \leq 7, \left(j = 1, 2, 3, 6\right) \tag{A.124}$$

$$x_j \geq 0 \text{ and integer} \left(j = 1, 2, \ldots 7\right)$$

A.3.11.3 Problem 3

$$\text{Minimise } f(x) = \sum_{j=1}^{j=7} x_j \tag{A.125}$$

$$\text{subject to } g(x_1) = x_1 + x_2 + x_3 \geq 6 \tag{A.126}$$

$$g(x_2) = x_4 + x_5 + 6x_6 \geq 8 \tag{A.127}$$

$$g(x_3) = x_1 x_6 + x_2 + 3x_5 \geq 7 \tag{A.128}$$

$$g(x_4) = 4x_2 x_7 + 3x_4 x_5 \geq 25 \tag{A.129}$$

$$g(x_5) = 3x_1 + 2x_3 + x_5 \geq 7 \tag{A.130}$$

$$g(x_6) = 4x_1 + 2x_3 + x_6x_7 \leq 15 \tag{A.131}$$

$$g(x_7) = x_4 \leq 15 \tag{A.132}$$

$$g(x_8) = x_5 \leq 15 \tag{A.133}$$

$$g(x_9) = x_j \leq 7, \; (j = 1,2,3,6) \tag{A.134}$$

$$x_j \geq 0 \text{ and integer} \, (j = 1,2,...7)$$

A.3.11.4 Problem 4

$$\text{Minimise } f(x) = \sum_{i=1}^{i=7}\sum_{j=1}^{j=7} x_i x_j - \sum_{j=1}^{j=7} x_j^2 \tag{A.135}$$

$$\text{subject to } g(x_1) = x_1 + x_2 + x_3 \geq 6 \tag{A.136}$$

$$g(x_2) = x_4 + x_5 + 6x_6 \geq 8 \tag{A.137}$$

$$g(x_3) = x_1x_6 + x_2 + 3x_5 \geq 7 \tag{A.138}$$

$$g(x_4) = 4x_2x_7 + 3x_4x_5 \geq 25 \tag{A.139}$$

$$g(x_5) = 3x_1 + 2x_3 + x_5 \geq 7 \tag{A.140}$$

$$g(x_6) = 4x_1 + 2x_3 + x_6x_7 \leq 15 \tag{A.141}$$

$$g(x_7) = x_4 \leq 15 \tag{A.142}$$

$$g(x_8) = x_5 \leq 15 \tag{A.143}$$

$$g(x_9) = x_j \leq 7, \; (j = 1,2,\ 3,\ 6) \tag{A.144}$$

$$x_j \geq 0 \text{ and integer} \, (j = 1,2,...7)$$

A.3.11.5 Problem 5

$$\text{Minimise } f(x) = \sum_{j=1}^{j=7} x_j^2 \tag{A.145}$$

$$\text{subject to } g(x_1) = x_1 + x_2 + x_3 \geq 6 \tag{A.146}$$

$$g(x_2) = x_4 + x_5 + 6x_6 \geq 8 \tag{A.147}$$

$$g(x_3) = x_1 x_6 + x_2 + 3x_5 \geq 7 \tag{A.148}$$

$$g(x_4) = 4x_2 x_7 + 3x_4 x_5 \geq 25 \tag{A.149}$$

$$g(x_5) = 3x_1 + 2x_3 + x_5 \geq 7 \tag{A.150}$$

$$g(x_6) = 4x_1 + 2x_3 + x_6 x_7 \leq 15 \tag{A.151}$$

$$g(x_7) = x_4 \leq 15 \tag{A.152}$$

$$g(x_8) = x_5 \leq 15 \tag{A.153}$$

$$g(x_9) = x_j \leq 7, \left(j = 1, 2, 3, 6\right)$$

$$x_j \geq 0 \text{ and integer} \left(j = 1, 2, \ldots 7\right)$$

A.3.11.6 Problem 6

$$\text{Minimise } f(x) = \sum_{i=1}^{i=7} \sum_{j=1}^{j=7} \sum_{k=1}^{k=2} x_i x_j x_k \tag{A.154}$$

$$\text{subject to } g(x_1) = x_1 + x_2 + x_3 \geq 6 \tag{A.155}$$

$$g(x_2) = x_4 + x_5 + 6x_6 \geq 8 \tag{A.156}$$

$$g(x_3) = x_1 x_6 + x_2 + 3x_5 \geq 7 \tag{A.157}$$

$$g(x_4) = 4x_2 x_7 + 3x_4 x_5 \geq 25 \tag{A.158}$$

$$g(x_5) = 3x_1 + 2x_3 + x_5 \geq 7 \tag{A.159}$$

$$g(x_6) = 4x_1 + 2x_3 + x_6 x_7 \le 15 \qquad (A.160)$$

$$g(x_7) = x_4 \le 15 \qquad (A.161)$$

$$g(x_8) = x_5 \le 15 \qquad (A.162)$$

$$g(x_9) = x_j \le 7, \ (j = 1, 2, 3, 6) \qquad (A.163)$$

$$x_j \ge 0 \text{ and integer} (j = 1, 2, \dots 7)$$

A.4 REAL-WORLD APPLICATION FROM MANUFACTURING DOMAIN

A.4.1 Multi Pass Turning Process

Nomenclature

C_I	machine idle cost due to loading and unloading operations and tool idle motion time $(\$ / piece)$
C_M	cutting cost by actual time in cut $(\$ / piece)$
C_R	tool replacement cost $(\$ / piece)$
C_T	tool cost $(\$ / piece)$
d_r, d_s	depth of cut for each pass of rough and finish machining (mm)
d_{r_i}	possible value of d_r under a given X
d_{rL}, d_{rU}	lower and upper bound of depth of cut in rough machining (mm)
k_3, k_4, k_5	lower and upper bound of depth of cut in finish machining (mm)
d_t	depth of material to be removed (mm)
D, L	diameter and length of work piece (mm)
f_r, f_s	feed rates in rough and finish machining (mm / rev)
f_{rL}, f_{rU}	lower and upper bound of feed rate in rough machining (mm / rev)
f_{sL}, f_{sU}	lower and upper bound of feed rate in finish machining (m / rev)
F_r, F_s	cutting forces during rough and finish machining (kgf)
F_U	maximum allowable cutting force (kgf)
h_1, h_2	constants relating to tool travel and approach/departure time(min)
k_o	direct labour cost + overhead $(\$ / min)$
k_t	cutting edge cost $(\$ / edge)$
k_1, μ, υ	constants of cutting force equation
k_2, τ, ϕ, δ	constants related to chip tool interface temperature equation
k_3, k_4, k_5	constants for roughing and finishing parameter relations $k_3, k_4, k_5 \ge 1$

n	number of rough cuts (an integer)
n_i	possible value of n under a given X
N_L, N_U	lower and upper bounds of n
p, q, r, C_o	constants of the tool-life equation
P_r, P_s	cutting power during rough and finish machining (kW)
P_U	maximum allowable cutting power (kW)
Q_r, Q_s	chip-tool interface rough and finish machining temperatures (8C)
Q_U	maximum allowable chip tool interface temperature (8C)
R	nose radius of cutting tool (mm)
SC	limit of stable region cutting constraint
SR	maximum allowable surface roughness (mm)
t_c	preparation time for loading/unloading (min)
t_e	tool exchange (min)
T, T_r, T_s	tool life, expected tool life for rough machining and expected tool life for finish machining (min)
T_p	tool life of weighted combination of T_r and T_s (min)
T_U, T_L	upper and lower bounds for tool life (min)
UC	unit production cost except material cost ($\$ / piece$)
V_r, V_s	cutting speeds in rough and finish machining (m / min)
V_{rL}, V_{rU}	lower and upper bound of cutting speed in rough machining (m / min)
V_{sL}, V_{sU}	lower and upper bound of cutting speed in finish machining (m / min)
X	machining parameter set $= \{x_1, x_2, \ldots, x_m\}$
θ	a weight for $T_p [0,1]$
λ, v	constants related to expression of stable cutting region

The multi-pass turning process problem is formulated based on the analysis given in Chen and Tsai (1996). The objective is to minimise the unit production cost by determining the optimal process parameters for the cutting speed, feed rate, depth of cut and number of rough cuts. The production cost UC for a multi-pass turning operation is divided into four basic costs:

1. cutting cost by actual time in cut, C_M;
2. machine idle cost due to loading and unloading operations and idling tool motion, C_I;
3. tool replacement cost, C_R; and
4. tool cost, C_T.

For the unit production cost the objection function is defined as the sum of the cutting cost C_M, machine idling cost C_I, tool replacement cost C_R and tool cost C_T.

$$UC = C_M + C_I + C_R + C_T \tag{A.164}$$

where

$$C_M = k_0 \left[\frac{\pi DL}{1000 V_r f_r} \left(\frac{d_t - d_s}{d_r} \right) + \frac{\pi DL}{1000 V_s f_s} \right] \tag{A.165}$$

$$C_I = k_0 \left[t_c + (h_1 L + h_2) \left(\frac{d_t - d_s}{d_r} + 1 \right) \right] \tag{A.166}$$

$$C_R = k_0 \frac{t_c}{T_P} \left[\frac{\pi DL}{1000 V_r f_r} \left(\frac{d_t - d_s}{d_r} \right) + \frac{\pi DL}{1000 V_s f_s} \right] \tag{A.167}$$

$$C_T = \frac{k_t}{T_P} \left[\frac{\pi DL}{1000 V_r f_r} \left(\frac{d_t - d_s}{d_r} \right) + \frac{\pi DL}{1000 V_s f_s} \right] \tag{A.168}$$

The problem is subjected to following constraints:

i)   The cutting force for rough and finish cut should be lower than the allowable force Fu

$$g_1 = k_1 f_r^{\mu} d_r^{\upsilon} - F_U \leq 0 \tag{A.169}$$

$$g_2 = k_1 f_s^{\mu} d_s^{\upsilon} - F_U \leq 0 \tag{A.170}$$

ii) The power for rough and finish cut should be lower than the allowable power P

$$g_3 = \frac{k_1 f_r^{\mu} d_r^{\upsilon} V_r}{6120 \eta} - PU \leq 0 \tag{A.171}$$

$$g_4 = \frac{k_1 f_s^{\mu} d_s^{\upsilon} V_s}{6120 \eta} - PU \leq 0 \tag{A.172}$$

iii) The stable cutting region constraints for rough and finish cuts are expressed as follows:

$$g_5 = SC - V_r^\lambda f_r d_r^\nu \le 0 \tag{A.173}$$

$$g_6 = SC - V_s^\lambda f_s d_s^\nu \le 0 \tag{A.174}$$

iv) The chip-tool interface temperature constraint for the rough and finish cuts are expressed as follows:

$$g_7 = k_2 V_s^\tau f_s^\phi d_s^\delta - Q_U \le 0 \tag{A.175}$$

$$g_8 = k_2 V_r^\tau f_r^\phi d_r^\delta - Q_U \le 0 \tag{A.176}$$

v) The surface finish constraint is defined as follows:

$$g_9 = \frac{f_s^\phi}{8R} - SRU \le 0 \tag{A.177}$$

The specifications and required operation parameters suggested by Chen and Tsai (1996) are in Table A.9.

A.4.2 Multi-pass Milling Process Problem

The multi-pass milling problem is adopted from Sonmez et al. (1999). This problem is solved to minimise the per unit production cost (i.e., maximise the production rate).

TABLE A.9 The Specifications and Required Operation Parameters.

$D = 50\ mm$	$L = 300\ mm$	$d_t = 6\ mm$	$V_{rU} = 500\ m/min$
$V_{rL} = 50\ m/min$	$f_{rU} = 0.9$	$f_{rL} = 0.1\ mm/rev$	$d_{rU} = 3$
$d_{rL} = 1\ mm$	$V_{sU} = 500$	$V_{sL} = 50\ m/min$	$f_{sU} = 0.9$
$f_{sL} = 0.1\ mm/rev$	$d_{sU} = 3$	$d_{sL} = 1\ mm$	$p = 5$
$q = 1.75$	$r = 0.75$	$k_1 = 108$	$\mu = 0.75$
$\nu = -1$	$\eta = 0.85$	$\lambda = 2$	$\upsilon = 0.95$
$k_2 = 132$	$\tau = 0.4$	$\phi = 0.2$	$\delta = 0.105$
$R = 1.2$	$k_0 = 0.5$	$C_0 = 6e11$	$h_1 = 7e\text{-}4$
$h_2 = 0.3$	$T_L = 25\ min$	$t_c = 0.75$	$t_e = 1.5$
$P_U = 200$	$T_U = 45\ min$	$F_U = 50$	$SC = 140$
$SR_U = 10\mu m$	$Q_U = 1000°C$	$k_3 = 1$	$k_4 = 2.5$
$k_5 = 1$	$k_t = 2.5$		

Nomenclature

a	Depth of cut for its pass (mm)
a_{max}	Maximum depth cut for machine tool workpiece system
a_{min}	Minimum depth cut for machine tool workpiece system
a_T	Total depth of cut (mm)
$b_v b_z$	Exponents determined empirically
B	Milling width (mm)
B_m	Correction coefficient of tool life equation
B_t	Correction coefficient of tool life equation
B_h	Correction coefficient of tool life equation
B_p	Correction coefficient of tool life equation
C_v	A constant taking into account the influence of all factors that are appearing separately in the tool life formula
C_{zp}	Constant of the cutting force equation
d_a	Arbor diameter (mm)
D	Outer diameter of the cutter (mm)
e	Permissible values of arbor deflection (mm)
e_v, e_z	Exponents determined empirically
E	Modulus of elasticity of arbor material $(kg / mm\,2)$
E_z	Modulus of elasticity of stub arbor material (MPa)
f	Feed rate (mm / min)
f_z	Feed per tooth $(mm / tooth)$
F_c	Mean peripheral cutting force (kg)
F_d	Permissible force with regard to arbor deflection (kg)
F_s	Permissible force with regard to arbor strength (kg)
I_s	Moment of inertia of stub arbor (mm^4)
k_b	Permissible bending stress of the arbor material (kg / mm^2)
k_t	Permissible torsional stress of the arbor material (kg / mm^2)
L	Length of cut (mm)
L_a	Arbor length between supports (mm)
L_s	Length of stub arbor (mm)
m	Exponent determined empirically
n_v, n_z	Exponents determined empirically
N_b	Total number of components in the batch
N	Spindle speed (rpm)
N_p	Number of passes
N	Number of sections
q, q_v	Exponents determined empirically
P	Exponent determined empirically
P_c	Cutting power (kW)
P_m	Nominal motor power (kW)
r_v, r_z	Exponents determined empirically
T	Tool life (min)

T_p	Machine preparation time per component (*min*)
T_s	Set up time of the machine for a new batch (*min*)
T_L	Loading and unloading time (*min*)
T_a	Process adjusting time and quick return time
T_c	Tool changing time per component (*min*)
T_d	Time for changing a dull cutting edge or tool (*min*)
T_m	Machining time (*min*)
T_{pr}	Total production time per component (*min*)
u_v	Exponent determined empirically
u_z	Exponent determined empirically
V	Cutting speed (*m / min*)
z	Number of teeth on the cutter
η	Overall efficiency
δ	Permissible deflection of stub arbor at the end (*mm*)
λ_s	Cutting inclination angle

For the multi-pass milling process, the total production time T_{pr} comprises four types of times as follows:

1. Machine preparation time (T_p)

$$T_p = \frac{T_s}{N_b} \qquad (A.178)$$

2. Loading-Unloading time (T_L)
3. Process Adjusting and Quick return time (T_a)
4. Machining Time (T_m)
5. Tool changing time per component (T_c) which is given by: -

$$T_c = \frac{T_d T_m}{T} \qquad (A.179)$$

where T_d is time required for changing a dull tool and T is tool life.
For a single pass, the total production time T_{pr} is given by:

$$T_{pr} = T_p + T_L + T_a + T_m + T_c \qquad (A.180)$$

or

$$T_{pr} = \frac{T_s}{N_b} + T_L + T_a + T_m + T_d \frac{T_m}{T} \qquad (A.181)$$

For a multi-pass operation, (A.181) becomes:

$$T_{pr} = \frac{T_s}{N_b} + T_L + \sum_{i=1}^{N_p} \left(T_{c_i} + T_{m_i} + T_d \frac{T_{m_i}}{T} \right) \qquad (A.182)$$

where N_p = total no. of passes, i denotes i^{th} pass.

For a particular milling operation, the machining time T_m is given as $T_m = \dfrac{L}{f}$

$$\qquad (A.183)$$

where L = length of cut and f = feed rate $= f_z z N$

f_z = feed per tooth, z = no of teeth, N = spindle speed in *rpm*.

$$N = \frac{1000 \times V}{\pi D} \qquad (A.184)$$

where D = cutter diameter, V = cutting speed.
The tool life can be determined by using the formula:

$$T = \frac{C_v^{1/m} D^{b_v/m} \left(B_m B_h B_p B_t \right)^{1/m}}{V^{1/m} a^{a_v/m} f_z^{u_v/m} a_r^{r_v/m} Z^{n_v/m} \lambda_s^{q_v/m}} \qquad (A.185)$$

where a = depth of cut and

B_m, B_k, B_p, B_t = correction coefficients;
$m, e_v, u_v, r_v, n_v, q_v, b_v$ = exponents
C_v = process constant
λ_s = cutting inclination angle

On substituting Equations (7.22–7.24) into Eqn. (6.21), the objective function for the multi-pass milling operation is expressed as follows:

$$T_{pr} = \frac{T_s}{N_b} + T_L + N_p T_a$$

$$+ \sum_{i=1}^{N_p} \left(\frac{\pi DL}{f_{z_i} z \times 1000 \times V_i} + \frac{T_d \pi L V_i^{\frac{1}{m}-1} a_i^{ev/m} f_{z_i}^{\left(\frac{u_v}{m}-1\right)} a_r^{e_v/m} z^{(n_{v/m}-1)} \lambda_s^{q_v/m}}{1000 \times C_v^{1/m} D^{b_{v/m}-1} \times \left(B_m B_k B_p B_t\right)^{1/m}} \right)$$

(A.186)

For each pass (rough and finish), the following three constraints are considered in the optimisation model (Rao and Pawar, 2010).

Arbor Strength
The arbor is subjected to torsion from the action of resistance to cutting. Therefore, the selected values of the process parameters should ensure that the arbor is safe from a strength point of view.

$$F_s - F_c \geq 0$$ (A.187)

where
Mean Peripheral cutting forces (F_c) is

$$F_c = C_{z_p} a_r D^{b_z} a^{e_z} f_z^{u_z}$$ (A.188)

C_{z_p} is a process constant and b_z, e_z, u_z are the exponents.
Permissible force for arbor strength $F_s (kg)$:

$$F_s = \frac{.1 \times k_b \times d_a^3}{.08 L_a + .65 \sqrt{\left(.25 \times L_a\right)^2 + \left(.5 \alpha D\right)^2}}$$ (A.189)

where

k_b is a permissible bending strength of arbor,
d_a is the arbor diameter,
L_a is the arbor length between supports,

$$\alpha = \frac{K_b}{1.3 \times K_t}$$

k is the permissible torsional strength of arbor.

Arbor Deflection

The selected values of process parameters should be checked for arbor deflection as:

$$F_d - F_c \geq 0 \qquad (A.190)$$

where the permissible force for arbor deflection $F_d\,(kg)$ is

$$F_d = \frac{4 \times E \times e \times d_a^{\,4}}{L_a^3} \qquad (A.191)$$

where E = modulus of elasticity,

e = permissible value of arbor deflection ($e = 0.2\ mm$, for roughing operation; $e = 0.05\ mm$, for finishing operation)

Power constraint

The power required for the cutting operation should not exceed the effective power transmitted to the cutting point by the machine tool. This is ensured by the equation:

$$P_c - \frac{F_c \times V}{6120} \geq 0 \qquad (A.192)$$

where

P_c is a cutting power (kW)
P_m is a nominal motor power
η is an overall efficiency.

$$P_c = P_m \times \eta$$

The following are the process parameters and their bounds:

a) Feed per tooth

$$f_{z_{min}} \leq f_z \leq f_{z_{max}} \qquad (A.193)$$

$$f_{z_{min}} = \frac{f_{min}}{zN_{max}} = \frac{14}{8 \times 2000} = 0.000875 \qquad (A.194)$$

$$f_{z_{max}} = \frac{f_{max}}{zN_{max}} = \frac{900}{8 \times 31.5} = 3.571 \tag{A.195}$$

where $f_{z_{min}}$, $f_{z_{max}}$ are min and max spindle feed rate, respectively.

b) Cutting speed

$$V_{min} \leq V \leq V_{max} \tag{A.196}$$

$$V_{max} = \frac{\pi D N_{max}}{1000} = \frac{\pi \times 50 \times 2000}{1000} = 395.84 \tag{A.197}$$

$$V_{min} = \frac{\pi D N_{min}}{1000} = \frac{\pi \times 50 \times 31.5}{1000} = 6.234 \tag{A.198}$$

TABLE A.10   The Specifications and Required Operation Parameters.

P_m (Motor power)	5.5 kW	Tool change time T_c	5 min
η Efficiency	0.7	Process adjust and quick return time	0.1 $min/part$
		T_a	
Arbor dia. d_a	27 mm	Lot size N_b	100
Arbor length between supports L_a	210 mm	Cutting angle λ_s	30°
k_b	140 Mpa	B_m	1
k_t	120 Mpa	B_k	1
E (modulus of elasticity)	200 Gpa	B_p	0.8
Spindle speed range	31.5-2000 rpm	B_t	0.8
Feed rate range f	14-900 mm/min	m	0.33
Tool material	HSS	e_v	0.3
Tool dia.	63 mm	u_v	0.4
No. of teeth z	8	r_v	0.1
Material	Structural carbon C%: .6%	n_v	0.1
Tensile Strength	750 MPa	q_v	0
BHN	150	C_v	35.4
Length of cut L_a	160 mm	b_v	0.45
Width of cut a_r	50 mm	C_{z_p}	68.2
Depth of cut a	5 mm	b_z	−0.86
Load/Unload Time T_L	1.5 min	e_z	0.86
Set up time T_s	10 min	u_z	0.72

c)   Depth of cut

$$a_{min} \leq a \leq a_{max} \tag{A.199}$$

$$0.5 \leq a \leq 4$$

Considering the specifications and required operation parameters suggested by Sonmez et al. (1999) are in Table A.10.

REFERENCES

Arnout, S. (2011) 'International student competition in structural optimization', ISCSO, www.brightoptimizer.com

Bernardino, H.S., Barbosa, H.J.C. and Lemonge, A.C.C. (2007) 'A hybrid genetic algorithm for constrained optimization problems in mechanical engineering', *Proceedings of IEEE Congress Evolution Computations*, pp. 646–653.

Chen, M.C. and Tsai, D.M. (1996) 'A simulated annealing approach for optimization of multi-pass turning operations', *International Journal of Production Research*, Vol. 34, No. 10, pp. 2803–2825.

Cheng, M.Y. and Prayogo, D. (2014) 'Symbiotic organisms search: A new metaheuristic optimization algorithm', *Computers and Structures*, Vol. 139, pp. 98–112.

Coello, C.A.C. (2000) 'Use of a self-adaptive penalty approach for engineering optimization problems', *Computer in Industry*, Vol. 41, pp. 113–127.

Coello, C.A.C. and Cortes, N.C. (2004) 'Hybridizing a genetic algorithm with an artificial immune system for global optimization', *Engineering Optimization*, Vol. 36, No. 5, pp. 607–634.

Deb, K. and Goyal, M. (1996) 'A combined genetic adaptive search (GeneAS) for engineering design', *Computer Science and Informatics*, Vol. 26, pp. 30–45.

Deb, K. and Srinivasan, A. (2005) 'Innovization: Innovation of design principles through optimization', KanGAL Report No. 2005007.

Eberhart, R. and Kennedy, J. (1995) 'A new optimizer using particle swarm theory'. *MHS'95. Proceedings of the Sixth International Symposium on micro Machine and Human Science*, Nagoya, pp. 39–43.

Efren, M., Coello, C.A.C. and Ricardo, L. (2003) 'Engineering optimization using a simple evolutionary algorithm', *Proceedings of 15th International Conference on Tools with Artificial Intelligence*, pp. 149–156.

Gandomi, A.H., Yang, X-S. and Alavi, A.H. (2011) 'Mixed variable structural optimization using firefly algorithm', *Computers and Structures*, Vol. 89, Nos. 23–24, pp. 2325–2336.

Hasançebi, O. and Azad, S.K. (2015) 'Adaptive dimensional search: A new metaheuristic algorithm for discrete truss sizing optimization', *Computers and Structures*, Vol. 154, pp. 1–16.

He, Q. and Wang, L. (2007) 'An effective co-evolutionary particle swarm optimization for constrained engineering design problem', *Engineering Applications of Artificial Intelligence*, Vol. 20, No. 1, pp. 89–99.

Kaveh, A. and Talatahari (2009) 'A particle swarm ant colony optimization for truss structures with discrete variables', *Journal of Constructional Steel Research*, Vol. 65, pp. 1558–1568.

Kulkarni, A.J., Kale, I.R., Tai, K. and Azad, K.S. (2012) 'Discrete optimization of truss structure using probability collectives', *Proc. IEEE 12th International Conference of Hybrid Intelligence System*, pp. 225–230.

Kulkarni, A.J., Kale, I.R. and Tai, K. (2013) 'Probability collectives for solving truss structure problems', *Proc. 10th World Congress on Structural and Multidisciplinary Optimization*.

Kulkarni, A.J., Kale I.R. and Tai, K. (2016) 'Probability collectives for solving discrete and mixed variable problems', *International Journal of Computer Aided Engineering and Technology*, Vol. 8 No. 4, pp. 325–361.

Lee, K.S., Geem, Z.W., Lee, S.H. and Bae, K.W. (2005) 'The harmony search heuristic algorithm for discrete structural optimization', *Engineering Optimization*, Vol. 37, No. 7, pp. 663–684.

Li, L.J., Huang, Z.B. and Liu, F. (2009) 'A heuristic particle swarm optimization method for truss structures with discrete variables', *Computers and Structures*, Vol. 87, Nos. 7–8, pp. 435–443.

Montes, E.M. and Ocana, B.H. (2009) 'Modified bacterial foraging optimization for engineering design', in Dagli, C.H. et al. (Eds), *Artificial Neural Networks in Engineering Conference (ANNIE), Intelligent Engineering Systems through Artificial Neural Networks,* Vol. 19, pp. 357–364.

Nanakorn, P. and Meesomklin, K. (2001) 'An adaptive penalty function in genetic algorithms for structural design optimization', *Computer and Structures*, Vol. 79, pp. 2527–2539.

Rao, R.V. and Pawar, P.J. (2010) 'Parameter optimization of a multi-pass milling process using non-traditional optimization algorithms', *Applied Soft Computing*, Vol. 10, pp. 445–456.

Rudolph, S. and Schmidt, J. (2012) 'International student competition in structural optimization', ISCSO, www.brightoptimizer.com

Sandgren, E. (1990) 'Nonlinear integer and discrete programming in mechanical design optimization', *Journal of Mechanical Design*, Vol. 112, No. 2, pp. 223–229.

Shih, C.J. and Yang, Y.C. (2002) 'Generalized Hopfield network based structural optimization using sequential unconstrained minimization technique with additional penalty strategy', *Advances in Engineering Software*, Vol. 33, Nos. 7–10, pp. 721–729.

Sonmez, A.I., Baykasoglu, A., Dereli, T. and Filiz, I´.H. (1999) 'Dynamic optimization of multipass milling operations via geometric programming', *International Journal of Machine Tools and Manufacture*, Vol. 39, pp. 297–320.

Sonmez, M. (2011) 'Artificial bee colony algorithm for optimization of truss structures', *Applied Soft Computing*, Vol. 11, No. 2, pp. 2406–2418.

Thanedar, P.B. and Vanderplaats, G.N. (1995) 'Survey of discrete variable optimization for structural design', *Journal of Structural Engineering*, Vol. 121, No. 2, pp. 301–306.

Tsai, Jung-Fa, Li, Han-Lin and Hu, Nian-Ze (2002) 'Global optimization for signomial discrete programming problems in engineering design', *Engineering Optimization*, Vol. 34, No. 6, pp. 613–622.

Wu, S.J. and Chow, P.T. (1995) 'Steady-state genetic algorithms for discrete optimization of trusses', *Computers and Structures*, Vol. 56, No. 6, pp. 979–991.

Yun, Y.S. (2005) 'Study on Adaptive Hybrid Genetic Algorithm and its Applications to Engineering Design Problems', MSc thesis, Waseda University.

Index

For Product Safety Concerns and Information please contact our EU
representative GPSR@taylorandfrancis.com
Taylor & Francis Verlag GmbH, Kaufingerstraße 24, 80331 München, Germany

For Product Safety Concerns and Information please contact our EU
representative GPSR@taylorandfrancis.com
Taylor & Francis Verlag GmbH, Kaufingerstraße 24, 80331 München, Germany